Power, Crisis, and Education for Liberation

Power, Crisis, and Education for Liberation

Rethinking Critical Pedagogy

Noah De Lissovoy

POWER, CRISIS, AND EDUCATION FOR LIBERATION
Copyright © Noah De Lissovoy, 2008.
All rights reserved. No part of this book may be used or reproduced in any manner whatsoever without written permission except in the case of brief quotations embodied in critical articles or reviews.

First published in 2008 by
PALGRAVE MACMILLAN™
175 Fifth Avenue, New York, N.Y. 10010 and
Houndmills, Basingstoke, Hampshire, England RG21 6XS.
Companies and representatives throughout the world.

PALGRAVE MACMILLAN is the global academic imprint of the Palgrave Macmillan division of St. Martin's Press, LLC and of Palgrave Macmillan Ltd. Macmillan® is a registered trademark in the United States, United Kingdom and other countries. Palgrave is a registered trademark in the European Union and other countries.

ISBN-13: 978-0-230-60275-5
ISBN-10: 0-230-60275-4

Library of Congress Cataloging-in-Publication Data is available from the Library of Congress.

A catalogue record of the book is available from the British Library.

Design by Scribe Inc.

First edition: June 2008

10 9 8 7 6 5 4 3 2 1

Printed in the United States of America.

Grateful acknowledgment is made for permission to revise and reprint the following essays, all by Noah De Lissovoy: "Frantz Fanon and a Materialist Critical Pedagogy" in *Critical Pedagogy: Where Are We Now?* edited by Peter McLaren and Joe L. Kincheloe (New York: Peter Lang, 2007); "History, Histories, or Historicity? The Time of Educational Liberation in the Age of Empire," *Review of Education, Pedagogy, and Cultural Studies*, 29(5), 2007; "Conceptualizing Oppression in Educational Theory: Toward a Compound Standpoint," *Cultural Studies ↔ Critical Methodologies*, 8(1), 2008.

For Arcelia and Paloma

Contents

Acknowledgments		ix
Introduction		1
1	The Time of Educational Liberation in the Age of Empire	9
2	Stretched Dialectic: Starting From Frantz Fanon	29
3	Conceptualizing Oppression in Educational Theory: Toward a Compound Standpoint	49
4	Clearings and Enclosures: Primitive Accumulation and Contemporary Schooling	79
5	Difference, Power, and Pedagogy	105
6	A Contemporary Philosophy of Praxis	129
7	Globality, Globalization, and Critical Pedagogy	157
Notes		179
References		183
Index		195

Acknowledgments

I am indebted to Peter McLaren for his support, mentorship, and friendship. His work and commitment serve as a constant example. This volume is deeply informed by our conversations and collaborations. I also wish to extend my thanks to Sandra Harding for her indispensable guidance, her careful reading of earlier versions of this book, and in general for her generous support. I am grateful as well to Kris Gutiérrez for her continual encouragement and helpful feedback. My initial engagement with critical pedagogy was nourished by my participation in the California Consortium for Critical Educators. I thank all my comrades from this group, and especially Antonia Darder, who convened it, and who has been a crucial mentor and friend ever since. In addition, I am thankful to all of my students, from whom I have learned so much. Finally, I am extremely grateful to my compañera, Arcelia Hernández, for the many discussions we have had about this project. This book is part of a larger journey. I give thanks to her for walking that path with me, and for showing me so much of the way.

Introduction

The idea of education as a liberatory practice is at once a commonplace within progressive educational thought and also, it seems, an increasingly improbable notion. On the one hand, once-radical theories of the political nature of educational processes have been absorbed into the body of work that defines the field and suggests problems for scholars and practitioners to explore. On the other hand, the idea that education can and should aim at a fundamental transformation of social relations, and that the vocation of educators can only properly be understood in the context of such struggle, seems to many to be more and more remote and problematic. To begin with, the apparently unobstructed operation of power within society and educational institutions tends to overwhelm the hope and vision that would contest it. In addition, as more of its hidden contradictions and complexities have been brought to light, "liberation" itself has sometimes come to seem somewhat suspect as a guiding framework. It has been pointed out that radical reforms in schools and other contexts have often been co-opted, or have themselves repeated the oppressions they aimed at eliminating. In addition, theorists have demonstrated how each notion of liberation, in the very moment of its articulation, represents in itself the operation of power and so silences at the same time as it empowers.

However, we are confronted in the present with continuing and deepening processes of domination and exploitation. In education, there have been dramatic efforts to restructure schools in reactionary ways. In politics more broadly, we see an ideological hardening and aggressiveness among elites. On a world scale, we are awakening to life under an empire that has reasserted itself in a way that we have only begun to apprehend and under an economic system that has given the lie to any hope of compromise and instead accelerated the drive to subsume all social and productive potential into the logic of accumulation. In the face of these processes, the left in general is in crisis, and so too those on the left in education. Given the depth of inequality in society and schools, a project that aims at anything beyond the protection of current minimums and prevention of further assaults can appear irresponsible, and an effort focused on understanding and overcoming enveloping social logics might seem grandiose. But if sweeping visions sometimes seem out of reach against these intractable processes, analyses that are restricted to local contexts are inadequate to the scope of the problem. If against what appear to be overwhelming challenges we abandon a comprehensive mapping of oppression and narrow the meaning of the idea of liberation to what can be achieved in local instances, we may be able to find some reassurance in the short term, but we do not solve the central problems. A *relative* liberation cannot be accomplished in the relative lessening of the degree of oppression. Liberation represents a qualitative change, a remaking of the order and intelligibility of a situation, not merely a bettering of it.

The dilemmas discussed above are compounded by a dramatic shift in the organization of social life that is currently taking place. This is the shift to a global horizon for political and economic processes, commonly referred to in the term *globalization*. Understanding this development means more than attending to the initiatives of neoliberalism or to the resurgence of imperialist practices, though this is an indispensable starting point. In addition, critical theory and pedagogy must be able to think through the existential challenge that the reality of the global forces upon all of us, as familiar identities and understandings no longer seem adequate to the scale of the transnational. In addition, at the level of strategy, the contest between hegemonic and counterhegemonic forces has to be rethought, since the mobility of capital makes it possible in many instances for power to evade decisive contests with oppositional movements. These movements now have the additional task of forcing the confrontation in the first place, and creating a space of struggle. This shift affects education as well, as schools are transformed from privileged instruments of ideological incorporation into just one more instance, on an increasingly homogeneous continuum, of a global social space emptied of meaning and possibility. I believe that we must refuse this state of affairs and that we must rediscover hope and faith in liberatory struggle; however, our hope must be informed by contemporary processes and conditions.

In analyzing this terrain, I argue that *occupation* has emerged as a crucial figure for understanding the operation of power and capital in the present. The crisis that we are living is of course not only the crisis of progressive movements, but first of all the fundamental crisis of capitalism itself. In this context, the imperative of accumulation means an increasing dependence by capital on projects of invasion and plunder in addition to familiar processes of exploitation (Harvey 2003). Politically, a resurgent imperialism results in literal occupations in the Middle East and elsewhere. Beyond this, the powerful strive to control the entire terrain of "democracy"—to capture the public space of dialogue for the simple language of force. Even more, capital seeks to invade the most intimate of experiences and to control the communicative and affective capabilities that make us human. The processes of invasion, occupation, and capture, then, are not merely names for military adventures, but also indicate the way that the time and space of the public, as well as the time and space of the self, are ever more deeply saturated by capital as a social logic. Importantly, this assimilation is at the same time a process of subjection to power. The daily violations by the state organize a spiritual plunder of people that is identified with the exploitation operated by properly economic processes. The prison-industrial system, for example, is the paradigmatic case of the profitability of repression, but on the other hand it also represents the reorganization of a mode of economic production—capitalism—as a system of social violence.

Education is one of the most important areas in which these tendencies are expressed. Not only does creeping privatization in this domain mirror the broader seizure of public space. In addition, there is a shift in schools away from the simple socialization of students to regimes of work and citizenship. The instrumentalization of curriculum, the proliferation of testing, and the ramification of new forms of hyperdiscipline seek to violate subjectivities as much as to produce them. Of course, the calculated demoralization of students can be seen as a strategy to police access to opportunities within the new global economy. But this is not the whole story, since the emptying of the space of the curriculum, as well as the authoritarianism of "zero-tolerance" discipline and supersecurity, mimic a more general carceralization of social life. As the borders of nations and neighborhoods are fortified, and as they are filled in their interiors with the blankness of a retreat into the private, schools too become both hardened and emptied. This is true both literally and figuratively: the metal detectors and security personnel are the outward manifestation of a broader repressive turn that also inflects pedagogical relationships, just as the inner vacuity of a test-based curriculum has its corollary in the literal emptying of students from schools through the instrument of the "accountability" system's demoralizing demonstrations of "failure." Given this context, critical pedagogy has to do more than champion dialogue in instruction. In addition, it should focus on interrogating the institution of the school as a whole as an instance of the operation of power in society more broadly, in the context not only of a critical curriculum but also of a social movement that extends beyond schools themselves.

The intensifying authoritarianism of educational culture calls for a more powerfully democratic conception of critical teaching and a reconsideration of the role of the teacher as leader. Even the critical authority that has been argued for as a goal for educators must be rethought in the context of a hardening of power that risks overdetermining consolidated senses of leadership as forms of domination. In this context, it is important to discover deeper possibilities for cooperation with students that decenter the teacher from the focus of instruction to crucial coparticipation. In the same way, it is important to reconceptualize learning and literacy as not only deeply distributive but also pointedly determined against the invasions of power into the terrain of knowledge. In the context of critical activism and cultural work more generally, the complex historical changes we are experiencing necessitate a collective project of interpretation and analysis. Critical pedagogy in this sense is deeply collaborative and can only be fully imagined in the context of broader movements. And in the practices of education and movement-building, as we think past familiar forms of authority and respond to the evasion by power of traditional terrains of contest, we must also engage in a profoundly original task, which is to set about building a new world even as we are captured by this one. In other words, even as we counter the assaults of the present, we need to see beyond them and start organizing a solidarity that begins to live beyond the logic of oppression. This project of construction requires in itself a more deeply democratic mode of engagement.

Liberatory struggle in general needs to be eclectic in the paradigms from which it starts and synthetic in combining approaches toward a more powerful analysis capable of conceptualizing the whole. For example, anticapitalist efforts focused on the protection of workers and opposition to the depredations of transnational corporations must be joined with struggles in defense of cultures and communities and against patriarchy and racism (Mohanty 2003). Only a compound standpoint anchored in the history of diverse struggles can respond to the convergence of oppressions that I describe above in terms of the figure of occupation. This political project needs to imagine liberation at the level of the social whole, understanding that this whole is not unified but complex. This analysis points to a more fundamental transformation than we have previously been able to conceive—transformation at the roots of the roots, at that incalculable location at which different exploitations and dominations find a common source. At this moment, such a concern for the totality does not turn out to be grandiose but is rather the necessary scale for defending the ultimately vulnerable *particularity* that this globe is, and the collective life that it supports. In this spirit, in this book I sketch the outlines of a new global oppositional class identification, which presses the notion of hybridity beyond itself and toward a fundamentally oppositional political determination.

If the emergence of the global means the emergence of a shared time across nations and communities, it also means an unraveling of the privileged history of the West, which has been projected as the singular time of the entire world. In this regard, globalization means the impinging on Eurocentric narratives of the perspectives of its global Others. But rather than resting in the moment of difference produced by a cosmopolitan conception of modernity, I argue in this volume that we need to start from the simultaneity of distinct global experiences and think through them toward a new shared political project. Fanon's (1963) injunction to the world outside Europe to discover a new nondominative humanism remains a crucial and as yet unfulfilled task for the world as a whole. We are called upon, I believe, to imagine a much vaster and more powerful solidarity than we have seen before. At the same time, honestly pursuing this solidarity means rethinking the forms of authority that activists and educators, from various points of privilege, have tended to reserve for themselves.

In this book, I make the argument outlined above by first reexamining the contributions of two crucial theorists. Chapter 1 considers Paulo Freire's work in light of contemporary political and cultural changes and reinvents his notion of "humanization" in order to respond to both the practical intensification of oppression as well as to the complexification of the idea of history in recent theory. In Chapter 2, I argue that the work of Frantz Fanon remains extremely useful for liberatory movements. In particular, Fanon can help us to rethink the principles of difference and dialogue, to understand the links between racism and class structuration within capitalism, and to imagine more democratic forms of teaching and activism. On the basis of the framework developed in these first two chapters, I argue in Chapter 3 that we need to join leading paradigms in critical education within a compound standpoint that is able to recognize the crucial links between different forms of oppression in education, and I describe how this synthesis can be theoretically anchored in an expanded understanding of materialism. Chapter 4 approaches the same problem from a different direction, analyzing recent concrete trends in school reform in terms of Marx's notion of *primitive accumulation*. From this perspective, accountability regimes, disciplinary processes, and privatization all participate in a process of enclosure that is at once economic and cultural.

The remaining chapters of the book propose a new oppositional subject in education and society. Beginning from a discussion of the idea of difference in critical educational approaches, Chapter 5 outlines a conception of cultural hybridity that is organized against capital as an encompassing global logic. By way of an analysis of the concepts of *oppressor* and *oppressed* in Freire, and *multitude* in Hardt and Negri, Chapter 6 describes a philosophy of praxis that can link struggles against contemporary forms of power to the project of building global democracy and proposes a new sense for class struggle that corresponds to this project. Finally, Chapter 7 confronts the problem of globalization head on, suggesting that this is as much a problem of the transition to a global frame of reference in social life as it is a problem of responding to the assaults of globalization as political economy. Understanding the different dimensions of globality that challenge critical educators and activists means rethinking the customary senses of identity that ground our work, as well as imagining more powerful forms of solidarity.

This book is rooted in my own experience as a teacher. I have learned from my students, at both the school and university level, that education is not essentially a formal or cognitive business but fundamentally a human situation. This has long been a basic principle of critical pedagogy, and however we respond to contemporary changes in society, this is a principle to hold to. I believe that for educators and activists, the discussion of power and liberation in this book points to the importance of keeping to a larger and more urgent vision of our work. Without losing the daily experiences of teaching and learning to some abstract concern for the grand scheme of things, we still need to refuse the framing of pedagogical questions as technical ones. We should remember that we are part of the struggle of ordinary people. To develop this struggle, we will have to think across the boundaries that restrict our main concerns as educators, and as intellectuals we will have to see past the disciplinary boundaries that attach us to our allotted niches. This is the true drama of the teacher: the stakes of pedagogical choices, as political choices, cannot be calculated in the narrow range of best practices, or even critical practices, or in the bland language of good "citizenship," but instead only on a political terrain in which we retain a sense of ourselves as *people* against the professional identifications within which society does its work with and through us.

The arguments and proposals I offer in what follows are necessarily limited by the particularity of my own perspective and experience. I hope that they will be useful enough to provoke supplements and improvements. Critical theory is effective to the extent that it contributes to broader struggles. And as they develop, so does their theory; the truth is not known beforehand. What I propose here is already the product of a collective life of critical teaching, thought, and action in which I have been lucky enough to participate. I hope that what I offer in this book can in some measure enrich this critical community and help to expand its resources in the face of the challenges that lie ahead.

CHAPTER 1

The Time of Educational Liberation in the Age of Empire

In approaches to education in the tradition of critical theory, the idea of history as a space of possibility allows for hope, intervention, and educational and political transformation. The process of historical becoming to which liberatory education is dedicated is both individually and collectively human. In this way, critical curriculum and pedagogy replace a narrow and disempowering notion of *history* (as the narrative of the actions of the powerful) with a dimensional, dynamic, unfolding notion of *historicity* (as the quality, for all of social life, of being situated in a determinate political moment). The historicity of life redefines history as a space of possibility that belongs to human beings as an ontological fact and an existential problem. The task of education is then to create the conditions in which students can act and intervene, as authentic subjects, in this historical situation. Paulo Freire, whose work is the focus of my discussion in this chapter, has importantly articulated this task in terms of the historical and educational vocation of *humanization*.[1]

However, recent trends in theory, practical politics, and schools themselves pose some important dilemmas for the conceptualization of history that is described above. Theorists have raised a set of problems with regard to the cultural standpoint from which the notion of history as development—as a narrative of dialectical *progress* through stages—has been elaborated. In addition, quickly changing conditions in practical politics and educational processes alter the possibilities for critical engagement and challenge our notion of the shape of historical time that underlies these conditions. Starting from work in contemporary postcolonial historiography and political theory, I discuss the most important of these issues, how they can be responded to by critical educators, and the implications for teaching contexts. While I argue in this chapter that in some important respects the work of Freire must be revised in the present context, it nevertheless remains a crucial starting point for educators and activists, and in its combination of political commitment and theoretical flexibility provides essential resources for the reconceptualizations of power, oppression, and liberation proposed in this book.[2]

Liberation and Historicity

Freire's conceptualization of the relationship between liberation and history and the pedagogical methodology that he proposes in response have been enormously influential among intellectuals, teachers, and cultural workers. His key texts continue to point the way forward for committed activists and educators. Nevertheless, he himself insisted that his work should be reinvented in different contexts and different moments. Ironically, in this regard he has often not been taken at his word even among proponents of his ideas. I suggest that the present moment requires a very particular reinvention of Freire, given the intensification of oppression and the rearticulation of its discourses that I analyze here. This reinvention extends to an interrogation of some of the key assumptions in the Freirean tradition. Nevertheless, I believe that such an analysis and the kinds of reconfigured praxis that can emerge from it confirm this tradition as foundational for contemporary liberatory efforts in education.

Freire's accomplishment in *Pedagogy of the Oppressed* and other works in which he explains the process of problem-posing education is not only to propose an educational approach but also to develop the notion of historicity in which this methodology is grounded (see also, in particular, Freire 1978, 1999). This space of history as possibility, which the rhetorical structure of his texts calls forth, is what allows for hope and for revolutionary futurity. In unfolding this terrain of possibility, which usually remains curled up in the apparently finished and inert "realities" of social life, Freire unfolds the dimensionality in which liberatory praxis becomes political reality. To become a part of this praxis is to participate in the creation of a world, and it is first of all the outlines of this world-possibility that his texts give us.

The space that is opened up here is not just a more expansive externality, in which human beings might more freely think and act, but the space of humanity itself. Humans are essentially historical beings, for Freire, and history is essentially human: "It is as transforming and creative beings that humans, in their permanent relations with reality, produce not only material goods—tangible objects—but also social institutions, ideas, and concepts. Through their continuing praxis, men and women simultaneously create history and become historical-social beings" (*PO*, 82). The process of *becoming* to which liberatory education is dedicated is at once the becoming of the world, the becoming of history, and the becoming of humanity. The sense of historicity as possibility that Freire evokes belongs to human beings, individually and collectively, as an existential problem. And the emergence of human beings as authentic historical subjects is at the same time the emergence of the world itself, as a world that comes to be real and complete through its construction and recognition by human consciousness, since "the world which brings consciousness into existence becomes the world *of* that consciousness" (*PO*, 63). This understanding is deeply Hegelian, and Freire's overall story is already given in the main from this tradition—deeply influenced in particular by the Lukács of *History and Class Consciousness* (1971). In Freire, however, this narrative is made concrete not only through a Marxist reconceptualization of history as essentially material and political, but also through an elucidation of the processual and pedagogical dimension of human development. The tremendous power of his work is that it offers not just a narrative but also a methodology of liberation.

The fulfillment of the historical vocation of humanization by the oppressed, and the overcoming of "the distortion of [this] call," involves a process of reflection and intervention in the world as problem and as possibility (Freire 1999, 98). This means participating in reality as process, as the making of a revolutionary future rather than the accumulation of a settled past. Reality is not a space of pure freedom, as bourgeois mythology would have it, but the space of a *situation*: a systematic set of problems and constraints, belonging to a determinate social context, that it is necessary to begin to see and to move against. The notion of reality as process contains more than the idea of being unfinished. It also includes the senses of human growing, transformation, and change through stages (Freire 1998). And further, to the extent that this transformation expresses a kind of reason, and is meaningfully, humanly organized as a process of *development* rather than an automatic or purely animal growth, then it must be a process of learning—it must be discovered and taught rather than simply manifested. Development is a "movement of search and creativity" that occurs in the "existential time of the conscious searcher" (*PO*, 142). In an interesting way, we might say that for Freire history itself is not only essentially human but essentially pedagogical as well. History is a human learning and a human teaching toward liberation. The centrality of pedagogy in this vision recovers two important aspects of history that are missing from a focus on consciousness by itself: history as a matter of relationship, dialogue, and sociality; and history as embodied human activity. It is no accident that these aspects are central to a Marxist understanding of history, which crucially informs his work; Freire, however, emphasizes them with the very particular intention of an educator.

Hope in Freire is not indeterminate but essentially political, as the hope of a revolutionary class and a faith that stands against oppression and exploitation. It is real and concrete precisely because it is attached to an apparently objective social logic. Here also, Freire's work is grounded in the Marxist tradition, in its faith in an immanent relationship between history and liberation. For Freire, at a moment in the dialectic of history that is defined by the domination of one class by another, and thus by the alienation of society from its own human essence, liberation represents the progression of this dialectic and the resolution of this estrangement. Liberation represents history's reconciliation with itself. In this way it can never be derided as the agitation of outsiders, or as unrealistic, since in a view taken of the whole, liberation is the central problem of history and the very condition of reality as a human category. And in this struggle the oppressed are not outsiders but rather the central protagonists: "The truth is, however, that the oppressed are not 'marginals,' are not people living 'outside' society. They have always been 'inside'—inside the structure which made them 'beings for others'" (*PO*, 55). In the terms of Marxist political economy, the mass of oppressed human beings, as exploited workers, are themselves the producers of the totality of social wealth and social power, even though "the combination of this labor appears . . . subservient to and led by an alien will and an alien intelligence—having its animating unity elsewhere" (Marx 1973, 470). The force, grandeur, and apparent historical centrality of the powerful, then, are only the effect of the labor of the exploited, a result of their human and creative capacity—and thus, in a large sense, even this alienated power is finally theirs, though it has been taken. Even as the oppressed, then, exploited human beings are the makers of history—but in revolution, which represents the resolution of the dialectic, they begin to make this history *for themselves*.

Cheating the Dialectic: Challenges to Historicism

For those who are committed to political struggle in the field of education, Freire's work is a powerful resource. What it communicates is not an abstract hopefulness, but a concrete and historical hope—not as a private disposition, but as a collective condition, the product of an actual passage of political struggle and transformation.[3] The reader does not take an artificial excitement from it; instead, the text restores her to her own real situation as a person in the world—as a political and human being in context, and as an actor and producer in and of that context. Nevertheless, the current juncture proposes a set of problems for the idea of history that underlies Freire's vision. First, developments in cultural studies and political theory raise a set of issues with regard to the discursive standpoint from which this notion of history is constructed. Second, a reconfiguration of liberatory possibilities both in practical politics and education policy implies a reconfiguration of our sense of the temporality they produce and proceed from.

Who Is History For?

Freire's conception of history is an optimistic one to the extent that it moves essentially forward, along the path of humanization, toward a radical emancipation for which we are "programmed," if not determined (1999, 98). This conception does not assume that the actual course of this development will be untroubled, but it does posit a particular directionality and historical telos as essential. But can we be sure that human history is really headed in one direction and that it will result in (or at least tends toward) the resolution, more or less, of the fundamental contradictions between subject and object, oppressor and oppressed, that constitute the "distortions" of dehumanization? More pointedly, does such a faith partly naturalize, in a roundabout way, these contradictions (and forms of oppression) as one stop along the road to reconciliation, as *necessary* to the constitution of this narrative of redemption? It is this sense that has led some—in particular, Adorno—to criticize the Hegelian dialectic as potentially totalitarian.[4] Similarly, if history is essentially the history of humanization, and if this development depends for its movement on the liberation of the oppressed—i.e., their emergence as historical subjects who can free themselves and their oppressors as well—then liberation becomes no less than "the people's vocation" or "the great humanistic and historical task of the oppressed" (*PO*, 25–26). But then *this* responsibility can seem to be "given" in much the same way that the mystifying immediacies of social reality initially appear to be "given." (It is after all this quality in the ideological apprehension of the world that Freire's work aims to dissolve.) Is liberation essentially *for* the oppressed, or for History itself, as unfolding dialectic? There are answers to these contradictions in dialectical reasoning, but at a basic level, the sense of history as a job to do doesn't always fit well alongside the sense of it as a space of radical possibility. Furthermore, if this History is defined essentially by domination and exploitation, then it could be argued that no ultimate redemption of it is possible, only an absolute overcoming of its total logic, value, and time (Negri 1984).

The problem, fundamentally, is *whose* story the world is assimilated to in this narrative of history and liberation as unilinear development, and as the progressive overcoming of the essential contradictions that drive this development forward. It should be noted that Freire makes a sharp distinction between development and modernization as the latter is understood by imperialist and bourgeois elites: "If we consider society as a being, it is obvious that only a society which is a 'being for itself' can develop. Societies which are dual, 'reflex,' invaded, and dependent on the metropolitan society cannot develop because they are alienated; their political, economic, and cultural decision-making power is located outside themselves, in the invader society" (*PO*, 142). Similarly, at the level of social psychology, Freire condemns the way the powerful pathologize oppressed people as backward and deviant, as beings of pure dependency. Against this, he argues that pedagogy must begin from an understanding of the oppressed as central, effective, and knowledgeable. And yet, he understands both Third World societies and the masses within them as submerged in a mystifying immediacy, and the story that he centrally tells is of their *emergence* and development. There must then be some vantage point—what Said has called "an invisible point of super-objective perspective" (1993, 167)—external to the oppressed themselves, from which this development is measured and recognized as participating in the authentic and universal schema. But how is the projection of such a schema not a repetition, at a secondary remove, of the kind of imposition that defines the cultural invasion by elites that Freire's work dissects?

This is the problem of Eurocentrism, as the epistemological ground and rhetorical organization of the European left philosophical tradition in which *Pedagogy of the Oppressed* in particular is deeply rooted. It is not clear that the notion of development, as compared with modernization, is any less of a return to the "stagist theory of history," in the terms of Chakrabarty, through which "European political and social thought made room for the political modernity of the subaltern classes" (2000, 9). Freire argues that third world societies are caught in an alien logic and time that petrifies them. And yet, his idea of the abstract authenticity of a developing society as a "being for itself" also seems to refuse that difference (between the universal time of Europe and *another* time of the "periphery") that colonialism seeks to destroy. According to Chakrabarty, "the achievement of political modernity in the third world could only take place through a contradictory relationship to European social and political thought" (9)—in which nationalist elites reproduced a Eurocentric historicism in their own praxis at the same time that they rejected Europeans' insistence on a "waiting-room" version of history, according to which the colonized were *not yet* ready for self-government. It may be that a similar contradiction, in another register, is at work in Freire. The emergence he evokes for the oppressed is supposed to be precisely *out of* the one-dimensional dialectic of (neo)colonialism. But this emergence itself, as a theory of history, seems to be at least partly produced by the assimilative and imperialist cultural logic it would defy. This is a familiar paradox for anticolonial movements, which have often ended up filling the places, in institutional terms, of their colonialist predecessors, as well as reproducing in their own discourse the Eurocentric grammar according to which postcolonial societies are understood as prepolitical and historically immature.[5]

Likewise, Freire's conception of the historical and educational function of dialogue may produce a similar set of effects. In her introduction to a volume of the work of the Subaltern Studies Collective of historiographers, Spivak writes that "the subaltern is necessarily the absolute limit of the place where history is narrativized into logic" (1988b, 16). It could be argued—against the sense in Freire of the adequacy of dialogue to its object and interlocutors, as the oppressed—that any notion of liberatory pedagogy needs to recognize that it can "never be continuous with the subaltern's situational and uneven entry into political . . . hegemony as the content of an after-the-fact description" (Spivak 1988b, 17). And the unevenness of this entry of the oppressed upsets the very continuity of the terrain of history. To emphasize this difficulty is, in a sense, to underline Freire's own concern about monological and culturally invasive forms of political practice that continue to assert an elitist view of cultural development and to dominate even in the name of "revolution." And yet, Spivak points here to a contradiction—between the historicist discourse of liberation and the actual complexity of politics—that organizes the notion of problem-posing education in Freire at its enunciation and that cannot be solved by a more scrupulous attention to consistency between word and action.

Behind Enemy Lines

Furthermore, at present, the conception of history that underlies Freire's narrative of humanization seems harder to hold on to in the face of concrete political developments. This history, with its leading edge pointing always toward emancipation, seems to darken in the present, to become overcast and ominous. It appears to be consolidated as the property of the powerful and to be sharpened into an instrument of invasion rather than unfolding as a horizon of possibility. Horkheimer and Adorno write that "the absurdity of a state of affairs in which the power of the system over human beings increases with every step they take away from the power of nature denounces the reason of the reasonable society as obsolete" (2002, 30–31). But this system and its power now seem to refuse even this truncated reason which they denounce. History no longer presents itself as a situation and a problem but rather as invasion and occupation. This is not only because power appears to have encircled the "untested feasibilities" that the oppressed are supposed to discover through praxis; it is also the result of lapses in the reason of oppression itself. Hope in Freire's liberatory vision is connected to the axiom that oppression is not opaque but is itself governed by a discernible logic—i.e., it represents a systematic distortion of humanization and a specific form of exploitation and control. But power in the present resists this logic; it often appears to (purposefully) produce its own failures. In global projects of "democracy building," as well as in domestic reorganizations of education and social welfare, the total ambition of reactionary "reform" engineers its own chaos and crisis. In its spectacle and excess, the way that the violence of power *functions*, and the way that it is *for* the class or class fraction that drives it, is much more complex. In this context, the dialectics of oppression and humanization must be rethought. To the extent that twenty-first–century imperialism materializes in history a new kind of evil or "corruption," it can be thought of as an awful progression of the logic of contradiction that governs the struggle between oppressor and oppressed (Hardt and Negri 2000). At the same time, however, contemporary imperialist violence often appears to exceed the existing order of interests that defines this contradiction—producing, in its rituals of occupation, detention, and torture, an excess and instability that tend to undermine the continuity of the classic ideals of imperial power and peace.

For Freire, the time of education depends upon the dialogue between teacher and student; in a larger sense, historical time depends upon a corresponding transaction between oppressor and oppressed—not in terms of dialogue, but in terms of struggle. But what happens when one antagonist seems to be evacuated from this dialectical relationship—to refuse to play by the rules? As Bauman (2000) points out, power in the present is not merely fortified but removed—placed at a maximum distance from popular forces. It preempts the possibility of any encounter within the space of communicative rationality that Habermas (1984) describes, and thus from any responsibility even to the language-game of liberal politics. The rationality that it projects to the public is less a form of propaganda than the trace of its disappearance from political space, a trail of debris that corresponds to the leftover bomblets that cover the territories of its interventions. In this moment, what becomes of the long "war of position" that was supposed to be fought in civil society and culture, and through layers of the state, to which Freire's program of collective conscientization and cultural revolution corresponds? This war has not been won by the rulers, but it does appear to have been overtaken by their dual strategy of abandonment and encirclement, as for example in the proliferating gang injunctions that make human association itself illegal and that seek to clear from territories the very possibility of human meaning—even oppressed and alienated meaning. In this context, the oppressed are not just dominated, according to the dynamics that Freire describes, but permanently occupied. In the words of Dead Prez (2000), Black people are collectively caught "behind enemy lines": "Standard routine—they put us in a box, just like our life on the block."

Furthermore, the quasi-legalization of torture, the invention of "enemy combatants," the Camp X-Rays and Deltas—these represent a festival of power for those who rule, the organization of its excess into a new intensity of application and experimentation. This intensification is different from the necrophilia that organizes the culture of oppression as Freire describes it in *Pedagogy of the Oppressed*. Death is not the object; the object is rather continual humiliation as the proof and image of absolute vulnerability and subjection. The point is not to conquer but to conquer *again and again*—conquest not as a drive or tool but rather as repeated play or reflex. In the historical process of alienation described by Freire, in their adherence to the oppressor and through the alienation of their own creativity and agency, the oppressed are reduced to objects passively acted upon by history. But the "elusive" enemies that the "war on terror" invents and "captures" are made *less* than objects. Denied even the ontological boundary allowed to the mute or petrified, power seeks to decide these real human beings as pure penetrabilities—as simple blanknesses that allow for the organization of "interrogation."

Disappearing Spaces, Proliferating Scripts

This foreshortening of the temporality of critical engagement and the condition of historicity it is connected to are reproduced in the educational arena as well. Increasingly, the very opportunity for educators to create critical and dialogical spaces is threatened as national and local policy initiatives limit the freedom of teachers, as Giroux (2003) has recently described in detail. On the one hand, a positivistic sense of knowledge and learning, and a renewed authoritarianism, collaborate to reduce curriculum to a set of instructions designed to produce the same results in every case. Confronted with scripted, "teacher-proof" materials, creative teachers have to choose between subterfuge, demoralization, or capitulation. Furthermore, the demand for precision in regimented programs and predictable texts means that both teachers and students are less able to take risks in teaching and learning, and that in fact they are themselves increasingly viewed as dangerous to the proper delivery of instruction (Meyer 2002; Starnes 2000). And it is not only scripted curricula that crowd the classroom space. Assessment, in the form of standardized testing, demands a larger and larger share of instructional time. The narrow senses of learning in these assessments impoverish the curriculum and orient it exclusively towards the production of better student "performances." Standardized tests require standardized test preparation packages, of course, further imposing upon the time and texture of instruction, as McNeil (2000) has shown, as well as supporting a bloated test preparation and analysis industry (Miner 2004). Much discussion has been devoted to the short-sighted conceptions of education that inform these standardized regimes of curriculum and assessment. Not enough attention, however, has been given to the way they diminish the space for alternative forms of pedagogy, and thus how they are connected to the larger processes of occupation described above.[6] It may be that the essential purpose of such initiatives is not to standardize teaching per se but rather to make critical and dialogical forms of engagement impossible.

In this way, education confronts in its own sphere the aggressive expansion of the logic of capital that characterizes the current moment. This expansion affects almost every region of private and public life, reconfiguring the flow and nature of bodies, information, and culture. For example, McChesney (1999) dramatically describes the hypercommercialism in the United States that has reorganized both the media and the advertising industry, which have become less and less distinguishable. While educational curricula are not always advertising vehicles in this way (though the branding of instructional materials and media, as well as school spaces and activities more generally, is an important trend as well—see Molnar 2005), the logic of standardization itself represents a penetration into education of the logic of capital, as student learning is construed as a continuous output that can be transparently represented by means of an abstract and universal measure. In other words, the space of teaching and learning—as pedagogy, curriculum, cognition, and culture—is increasingly reified in a way that mirrors the more straightforward commodification of public space generally. In fact, Jameson (1991) suggests that one way that we can understand postmodernity is as the kind of global space that is created as a result of the colonization by capital of every enclave that we understood previously as its own and separate sphere. As this expansion approaches the absolute, it becomes harder to find any meaning for culture (and education as a subset of it) that is not immediately an expression of this predatory logic. The constraints that this places on critical dialogue in teaching, and the kinds of resistance that are necessary, are therefore deeply connected to the limits on democratic and transformational discourse and action in society more generally.

Thus, a dialogical and liberatory educational program in the present has to confront something *more than* hegemony, as it is traditionally understood. Apple's principle that for progressive educators "the basic act involves making the curriculum forms found in schools problematic so that their latent ideological content can be uncovered" (1990, 7) has to be rethought at this juncture. Before it is a question of the ideological construction of curriculum, it is a question of the possibility of curriculum in the first place, in the face of the reconfiguration of teaching as absolute procedure (De Lissovoy and McLaren 2006). And before it is a matter of conflicting understandings—of dominant and subordinate knowledges, for instance—it is first of all a matter of the almost complete monologue of official discourse, or rather of what might be called *hyperlogue*: the filling of public life and of the curriculum with an excessive script (as advertising, "news," testing, etc.) that seems to leave no room for even a hegemonic construction of knowledge (as a worked out and internally consistent system). In this context, the traditional critical emphasis on *voice* needs to be rearticulated, since the initial task is to see through the onslaught of utterances that already occupies the space of thinking and speaking.

Response and Reconceptualization

The difficulties that I have described above are, in the Freirean sense, problems as challenges. They do not imply that it is impossible to grasp the dynamic of history, or histories. Rather, considering these questions can make this understanding more real and more relevant. An adequate conception of history as the time of praxis is essential in order to be able to specify the ground of pedagogy. In responding to the issues raised, and in reconceptualizing the temporality of liberation, it is important both to recover the process of humanization against efforts to erase it and to complexify the idea of history with which it is associated.

Recovering History as Humanization

It is crucial not to naturalize the operation of contemporary oppression as if it belonged to an absolutely new and different order of power. *Against* the intensifications I have just described, it is important to read the links between contemporary processes and the larger history of exploitation and domination. In recognizing the particular circuits of violence in the present, we also need to recognize how they are connected to familiar economies. In particular, while the radical humanism of critical educational approaches needs to be rearticulated, it remains indispensable, since it is part of the purpose of contemporary invasions and occupations to erase that humanism's scene and memory. The assertion of a liberatory humanist commitment, albeit in the context of a less enchanted sense of its centrality in history, is more necessary than ever—in order both to expose the intensity of violence and to recover the story of resistance that lives and fights even in isolation, through the "immunization" that the oppressed body develops "to defend itself against the harsh conditions to which it is subjugated" (Freire and Macedo 1987, 137). If capital in the present seeks to absorb social space completely and to reduce completely the human beings that outrage the perfection of its alien logic, then witnessing those beings and testifying to their presence and value, even where layers of *information as silence* would try to convince us that they never lived, is an urgent political act.

One important way to recover a sense of history as the project of humanization is to return to the radical emphasis in the tradition of problem-posing education against co-optations and ill-advised critiques. Efforts to appropriate this methodology in order to ameliorate and "humanize" dominant educational cultures betray the purposes of this pedagogy. In this regard, it is important to emphasize a characteristic of Freire's account of dialogue that is generally overlooked and that he himself did not fully elaborate: dialogue *does not mediate* the most fundamental difference that traverses the social, the difference marked by oppression itself. Instead, dialogue is a process that is *internal* to the formation of a revolutionary class—in fact it is the crucial mode of organization of this class against oppression, and against the oppressors. For this reason, "dialogue between the former oppressors and the oppressed as antagonistic classes was not possible before the revolution; it continues to be impossible afterward" (*PO*, 120). Dialogue creates the space for the emergence of a subject that is able to intervene for itself in history. In this way, rather than mediating class conflict, dialogue is an organic form of empowerment among participants on one side in a necessary process of social contradiction and struggle.[7] Not only does dialogue challenge dominant educational paradigms by reconceptualizing teaching as collaborative production rather than as recitation, in this way shifting the center of gravity of education as an activity from the teacher to the space between teacher and students; in addition, dialogue is the essential mode of praxis against a form of power that depends upon the objectification of the oppressed, since "no oppressive order could permit the oppressed to begin to question: Why?" (*PO*, 67). Dialogical praxis is needed in order to challenge a dominative power that proceeds by way of dehumanization; at the same time, as educational methodology, dialogue confronts an authoritarianism in teaching that is identified with the broader logic of oppression in society. Recovering this determination of dialogue as radically antiauthoriarian is especially important in the context of the current intensification of structures and processes of control.

The political determination of dialogue has other important consequences. Most importantly, it means that the goal is not indifferent communication but rather political and spiritual intercommunication or *communion*. Unlike mainstream senses of dialogue, this conception is fundamentally partisan; it stands on the side of the oppressed against contemporary dehumanizations. And however novel its current forms appear, oppression in the present is grounded ultimately in the same bourgeois and dominative reason that Freire's methodology originally aimed to unravel. The distinction between his radical conception of dialogue and the discursive intersubjectivity that is the goal of dialogue in a liberal or Habermasian frame has to be sharply drawn. In this sense, it is necessary to return its original Buberian emphasis to the Freirean tradition; for Buber, dialogue is not essentially a matter of speech but rather of existential orientation—of human relation in the most fundamental sense: "The basic word I-You can be spoken only with one's whole being. The concentration and fusion into a whole being can never be accomplished by me, can never be accomplished without me. I require a You to become; becoming I, I say You" (1970, 62).

In critical education, the establishment of this relation must have a material and political valence that concretizes Buber's schema: the reciprocal recognition of whole and authentic human being has to be undertaken against an organization of social relationships that systematically violates and precludes it. This struggle is quite different from rationalistic communication or consensus-bringing argumentative speech. In fact, clearing a space for the possibility of human subjectivity and dialogue means opening a breach with the procedurally determined senses of communication that characterize many liberal conceptions. Dialogue *as praxis* cannot be safely redeemed according to universally acceptable norms—nor can the subjects who emerge within it, who are always unfinished, always still becoming. This does not mean that critical dialogue does not position its participants in specific ways, as some have argued (e.g., Ellsworth 1997). However, it is a misunderstanding of critical dialogue to say that it positions its participants as transparent or unproblematic subjects. In this sense, to continue to struggle for a notion of history as humanization is not to attempt to reassert a familiar and static subject of history and liberation but rather to insist on the open-ended and indeterminate space itself of the "human"—against its current enclosure and erasure.

Histories Against and Retrograde Motion

In the Freirean tradition, as the oppressed learn to challenge an encompassing and paralyzing reality, they affirm themselves as historical beings—as participants and producers of history as praxis. The process of liberation coincides with an awakening to historicity—a claiming of that time for the becoming of the mass of men and women. This is also a movement of the margins to the center—the discovery of their centrality by those who have been pathologized as "marginals." In the present context of power and capital as terror and occupation, is it possible to inflect the essential emphasis on the political agency of the oppressed slightly differently—i.e., to articulate this agency in relation to history in a way that responds to the specificity of the present?

Adorno invented a negative dialectics against a politics of affirmation in a society in which all positivities were already given by a totally administered reality. In the same way, as history itself is seemingly monopolized by the rulers (or at least as the image of this assimilation is more powerfully projected), participation in it, even as an official antagonist in struggle, begins to feel like an unacceptable level of collaboration with an alien time. In this context, global resistance points not to an empty space outside that History but rather to another history, *other histories* that both escape it and contest it: the concrete narratives of resistance by people who have suffered colonization and occupation. These narratives constitute different histories, or *histories against*, that are motivated not by the hope of finally dominating historicity itself but rather by the need to assert a rejected humanness, even when domination refuses to properly open to historical possibility. Audre Lorde writes:

> and when we speak we are afraid
> our words will not be heard
> nor welcomed
> but when we are silent
> we are still afraid.
> So it is better to speak
> remembering
> we were never meant to survive. (1978, 32)

This spirit of resistance as survival shares a great deal with Freire's radical humanism. But Lorde's refusal to submit, at any cost, to a present that is overwhelmingly violating of her own human difference and person is still different in emphasis from a faith in a single time of the world, as redemption, that ultimately opens up to take one inside it.

How can resistance in this sense, as the claiming of her own historical *outside* against erasure, be usefully brought together with Freire's faith in a *general* project of humanization?

It is also useful to juxtapose Freire's historical optimism with other dispositions toward history within Marxism. What happens when resistance must take place against all odds? Where do the resources for fighting come from when history withholds any guarantee, or when it seems to guarantee destruction? The larger question would be, can acts of revolutionary love be possible in the retrograde motion of uneven development, in which praxis might not coincide with *winning* at any remove? Bertolt Brecht's work suggests the virtue of a kind of stoicism, in the face of oppression, that is able to take realistic stock of the balance of forces. The rage of the violated does not by itself overwhelm injustice and cannot do so without regard to the objective situation.[8] There are moments when political wisdom recommends inaction. Likewise, James (1992) emphasizes the forward and backward movement of history, and the importance of recognizing the real catastrophes that have followed the failure of democratic and socialist aspirations. Our sense of history has to include, and essentially so, these collapses and the painstaking regroupments that follow. These moments of waiting, frustration, and stasis are not ruled out in Freire, but the idea of history in his work does not emphasize the eddies and chutes that make it something more than the space of forward movement.

Pedagogical Implications

The challenges to the idea of historicity that I have described above, as well as the possible responses I have elaborated, are deeply connected to the terrain of pedagogy. To the extent that Freire's work is premised on a particular conception of the relationship between history and liberation, and to the extent that education is inherently involved in the idea of history as a human problem, then the discussion above has important implications for teaching and for teachers and suggests some key principles for problem-posing pedagogy in the present.

Critical education takes the settled facts and truths of conventional education (and history itself) and proposes them to students as objects to be investigated, interrogated, and *un*settled. It poses them as *problems* in relation to students' own lives. An important question, however, is whether the historical "situation" here includes the pedagogical encounter itself, i.e., the very moment of dialogue and reflection. Such a deepening, or infolding, of the sense of history and situation is necessary in the present. This is because as the dimensionality of historicity and the horizon of critical engagement are foreshortened, as I describe above, the essential generative themes that Freire's methodology aims to discover and explore with students inhabit more and more aggressively the very moment of teaching. The overdetermination of pedagogy by forms of control (mandated curricula, a reduction of learning to obsessive assessment, teaching as punishment, etc.) is immediately continuous with the reconfiguration of the conditions of history and of the operation of power in social life generally. This continuity calls for attention to the material and discursive conditions in which pedagogical dialogue takes place. Being attentive to these conditions does not mean being pessimistic about the possibility of problem-posing dialogue; rather, it means more acutely focusing the political reading of the *problem* as a totality—a totality that includes the moment in which critique itself has to occur.

In addition, the new forms of historical alienation that I describe in this chapter are connected to a new condition of globality, as the violence of global processes is no longer insulated from the regional by any buffering distance. Contemporary processes of globalization challenge the common-sense conceptual opposition between the "global" and the "local," as Sassen (1998) has described, and the common-sense identification of history only with what occurs on the dramatic and visible stage of the former. Paradoxically, intervention in history is then as much a matter of an *inward* and proximate attention as it is an orientation to the *forward* and *outward*. Here the "world," in the Freirean sense of the complex of encompassing historical contradictions, is felt very close up—as close as the walls and barred windows that surround and construct the space of teaching, as close as the complicated schedules that mark off a new array of educational experiences—new kinds of "learning," monitoring, down time, and punishment. These proliferating and yet porous forms of discipline, these new proceduralizations and hypercontrols—which are at the same time, in their excess, new forms of abandonment—are identified at once with the logics that reconfigure public space more generally (Devine 1996; Saltman and Gabbard 2003). Dominant progressive rhetorics of social justice in education tend to focus on how pedagogy can "promote or constrain the degree to which students acquire the knowledge, skills, and attitudes necessary to function effectively as citizens in a democracy" (Westheimer and Kahne 1998, 19). They generally ignore the fact that critical education in the present is charged with the larger task of making the connection between local, regional, and global processes visible. In this context, attending to the space of schooling itself is an essential moment in the total picturing of oppression and a precondition for the imagination of new kinds of solidarity that can connect students' experiences and forms of resistance to moments of opposition more generally.

Conclusion

Contemporary developments in theory and practical politics pose challenges for the notion of historicity that underlies the Freirean conception of liberation. The progressivist and teleological narrative of human development that informs this tradition are called into question by critiques of Eurocentric historiography. In addition, the simple faith in historical overcoming that is understood as the properly emancipatory attitude for educators and cultural workers toward history needs to be complexified through an understanding of the specificity and intensification of social and political alienation in the present and the scale of its operation. Nevertheless, such challenges only make more urgent the necessity for critical educational theory of rearticulating and recovering the project of humanization. This will involve developing a more sophisticated sense of history as both plural (i.e., as a matter of *histories* vs. *history*) and multidirectional, involving retrograde as well as forward movement. Educators might then base their work on these reconceptualized senses of history and humanization, in order to be able to intervene more effectively in the immediate moment of teaching, as well as to connect these interventions to larger political struggles. Against an overly static and universalized notion of the relationship between history and liberation, transformational love and humanism should inform a contemporary oppositional pedagogy not as abstract values, but instead as situated practices and relationships. From this place, concrete affiliations can be imagined with others, elsewhere, and in this way a kind of humanity can be woven that is no more than a spiritual and political community in action.

CHAPTER 2

Stretched Dialectic
Starting From Frantz Fanon

In U.S. schooling, the racist disparities in the allocation of resources and the cultural coding of standards and evaluation approach a kind of apartheid. Not only does a "second-generation" de facto segregation relegate most students of color to substandard facilities and instructional materials; in addition, public education is increasingly in the grip of a hyper-reductionistic "accountability" movement that impoverishes the curriculum while also constructing the performance of low-income students, through the tool of the standardized test, as failure. Increasingly militarized schools aggressively enroll youth of color and poor white students into global projects of conquest and empire building; at the same time, these imperialist adventures correspond to the almost complete abandonment by the state of its already limited commitments to its own constituents and to the basic social institutions that serve them.

In understanding these conditions and the colonial relationships that produce them, the work of Frantz Fanon (1925–1961) is a crucial starting point. Fanon was a brilliant psychiatrist and revolutionary theorist, originally from Martinique, who was active as a leader and intellectual in the Algerian independence movement. His work has been influential in resistance movements throughout Africa and the rest of the world and has also been indispensable in understanding the cultural and psychological dynamics of colonialism and racism, as well as decolonization. In contrast to the dominant ways in which social inequality is understood, Fanon's work analyzes cultural life as inextricably linked to politics and to histories of violence, power, and exploitation. Fanon demonstrates that these histories traverse a range of registers, from the physical and psychic to the cultural and economic, and he describes a project of resistance that comprises all of these dimensions. For critical educators, these insights are crucial resources for challenging the technicist framework of educational reform, the reification of race and culture, and their separation from material relationships of domination, as well as overly economistic analyses of social oppression.

In this chapter, I contrast Fanon's work with other critical analyses of oppression in society and culture, and consider the implications of his thought for critical pedagogy and liberatory educational projects. In particular, I describe how both Marxist and postmodernist approaches in educational theory can benefit from a consideration of Fanon's insights and the ways in which his work challenges presuppositions of both of these traditions. In the course of this discussion, I outline a "Fanonian" materialist perspective that links the cultural, political, and economic in a complexly dialectical conception. This understanding returns, I believe, to the original Marxist sense of the material as constituted by the historical organization of social relationships of production. This framework suggests a form of decentered humanism that is global in its commitments and revolutionary in its intentions, which I believe crucially extends the possibilities for critical pedagogy in the present.

Racism, Colonialism, and Dialectical Analysis

In Fanon's analysis of the historical processes of colonialism and anticolonial struggle, a Manichean opposition organizes colonial society into a rigid binary:

> The originality of the colonial context is that economic reality, inequality, and the immense difference of ways of life never come to mask the human realities. When you examine at close quarters the colonial context it is evident that what parcels out the world is to begin with the fact of belonging to or not belonging to a given race, a given species. In the colonies the economic substructure is also superstructure. The cause is the consequence; you are rich because you are white, you are white because you are rich. This is why Marxist analysis should

always be slightly stretched every time we have to do with the colonial problem. (1963, 40)

In this account, a dynamic of domination mediates the opposition between colonizer and colonized that differs from the naturalized order of exploitation within metropolitan capitalism. Economic position is identified first of all with racial assignation rather than being essentially an effect of social class. Fanon articulates here what he calls a "stretched" dialectic of oppression. There are several aspects of this analysis that are important. First of all, the central contradiction in classical Marxism between labor and capital is pulled apart to encompass a different one: the opposition between two "species," as Fanon puts it—black and white, the contradiction of colonial racism. This represents a displacement of the class contradiction; but in addition, the nature of the opposition that Fanon describes is different. The logic that mediates the relationship between colonizer and colonized is one of violence rather than incorporation. Colonial society is organized to begin with by the absolute separation of white and black, instead of by the intimate relationship, in production, that defines the contradiction between classes. Nevertheless, this is not some entirely different social universe from that described by classical Marxism; the fact of belonging to a particular race is still connected to political economy: "You are rich because you are white." And as in a Marxist frame, a central antagonism systematically produces social reality, even if it is not the antagonism between capital and labor.

Fanon's text produces an asymmetry, since it focuses the language of Marxist critique on a phenomenon usually thought of as analytically secondary in this theoretical tradition—namely, racism. This rhetorical asymmetry itself conveys the complexity of the colonial context. In its very deployment of the language of Marxism, Fanon's narrative ironically evokes its own distance from this methodology (as traditionally understood) and appropriates its own "distortedness" as oppositional and productive. In fact, this distortion of the Marxist dialectic is also a moment of its development and advance. Fanon teaches us the importance of critical readings that are caught in the tension between successfully recuperating their objects and registering the resistance of reality to this recuperation: an off-balance dialectic for a world that is itself also essentially off-balance.

This stretched dialectic can allow us to make sense of phenomena in education that appear incoherent from the perspective of a simple model of reproduction of class society (which has dominated critical sociology of education). Thus, power in schools operates not only to order, organize, and produce, but also to violate, refuse, and expel (Devine 1996). This power also operates differentially—socializing differently raced students differently. Zero-tolerance disciplinary policies, "tough love" retention schemes, and high school exit exams are all examples of popular educational initiatives designed to preserve advantages for white students through the partial or complete expulsion of the students of color that they disproportionately target (Dohrn 2001). These aspects of education appear anomalous within a system that is thought to be oriented toward the socialization and incorporation of students as disposers of labor power or as the functional citizen-subjects of bourgeois society. However, from Fanon's perspective these processes can be understood as systematic—the expression of an organized violence and colonialism. They should suggest to us that the essence of domination in education is represented as much in the *subjection* of students of color as through the ideological *interpellation* of students generally.

The relevance of Fanon's analysis to contemporary education is especially evident in the rapid militarization of schooling that is presently occurring. As a result of more and more aggressive recruiting tactics as well as the articulation between schools and the military that the No Child Left Behind Act makes automatic (in furnishing the names of high school students to recruiters as a matter of policy), students of color and poor white students are pressed with even greater energy into the service of neocolonialist projects (Furumoto 2005). While preparation for military service has arguably always been one function of schools, the contemporary institutionalization of recruitment as a central schooling function is new, and inflects in a new way our sense of the very idea of citizenship for which schools are taken to be a preparatory mechanism, as Henry Giroux (2003) describes in relation to patriotism more generally. This process of militarization—of the invasion and reorganization of the space of civil society by the war machine—reflects at some distance the actual imperialist invasions of the post-9/11 era and the destruction and distortion of the social that they have occasioned in Afghanistan and Iraq. These invasions are also powerful testimony to the continuing significance of the problematic of colonialism, as described by Fanon, in contemporary global society.

For Marxists, not to attend to the complexities Fanon describes is either to deny the basic salience of his account or to restrict the frame of reference of socialist theory to a Eurocentric framework. As Fanon points out, "the dialectical strengthening that occurs between the movement of liberation of the colonized peoples and the emancipatory struggle of the exploited working classes of the imperialist countries is sometimes neglected, and indeed forgotten" (1967b, 144). What is in order is precisely an extension of left analysis, an exploration of the relationships between different systems of social oppression. Such an exploration would not posit any arbitrary unity but would nevertheless persistently attempt to conceptualize the whole. Racism, in Fanon's account, produces a social order: colonialism. Likewise, while capitalist exploitation is usually thought of in terms of economic production, it is also a moment of social *violence* in creating "free" workers as such (Marx 1867/1976). The production of racial and class positions, their intersection and distinctiveness, must itself be carefully thought out, not in order to equalize or identify them but rather to begin to glimpse the total universe they point to together.

Rethinking Difference: A Materialist Framework

The idea of difference has been ubiquitous in recent theory, both in the academy generally and in education in particular. In philosophy, a preoccupation with difference has been one of the hallmarks of postmodernism; in education, it has been a crucial organizing principle of theories of multiculturalism. This emphasis has often been counterposed to an old-fashioned universalistic tendency to "synthesize . . . differences into a unitary, univocal whole" (Flax 1990, 4) and to imagine liberation and education in terms of a series of unifications: of the class, of the society, of the teacher and students. In this way, difference has been thought of as tracing the boundaries of a sequence of experiences, cultures, and identities that the modernist language of the left refused or erased. However, this conceptualization of difference is marked by a certain idealism, since it tends to understand differences as either marking static identities (e.g., "cultures"), or as infinitely variable, the space of the free play of language and subjectivity. On the other hand, the "old" left has sometimes been too happy to cede this conceptual terrain to the postmodernists as a secondary dimension that does not affect the underlying contradictions that determine social life. In Fanon, this false opposition is exploded; he demonstrates the dynamic and dialectical processes that difference is the product of, and the historic and violent process of differentiation that ties subjects always to the social totality. Cultures do not simply congeal into the categories that they are often assumed to be in educational discourse but are always the constructed and historical products of struggle, and their fates are inextricably bound up with the political aspirations of the people: "There is no other fight for culture which can develop apart from the popular struggle" (1963, 233). It is in this sense that Fanon's conception of difference can be understood as materialist, as opposed to the often purely discursive, symbolic, and ideal conceptions that predominate in much cultural and educational theory.

This understanding of the material dimensions of difference poses an implicit challenge even to forms of progressive education, such as multiculturalism, which seek to honor differences and to challenge inequities through a reframing of cultural valuations and affective relationships. The problem is that the very emphasis in these approaches on the emotional and ethical concern of teachers for their students tends to overlook the *economy* of caring itself—the materiality of relationships in education that are themselves caught up in a circulation that produces both scarcity and surplus. The emphasis on caring and expectations tends to admonish individual teachers to transform their relationships to students without recognizing the fact that this attention and concern is itself a key resource systematically and structurally distributed in favor of dominant groups. Paradoxically, caring does not merely describe an affective state of the teacher but in fact is already differentially attached to students as a kind of right—what Cheryl Harris (1995) describes more generally as a "property interest in whiteness." The disparity in concern cannot be fixed by merely switching the disposition of the teacher, since this relationship is not simply a matter of attitude but is already crucially determined by the differential valuation of students in the total practice and system that is schooling, and within which pedagogy is made coherent. Elaborating from Fanon, we might say that a schooling system, like a society, is either racist or it is not, and that all (neo)colonial systems are racist. It is not a matter of adjusting an attitude but rather of struggling against an overarching social logic and set of relationships. Even the category of love, in teaching, has to be understood materially; it is part of the structure of racism in education that in this libidinal economy, as in the dollar economy, resources are systematically plundered and redistributed across the social topography in ways that impoverish learning spaces for students of color.

However, postmodernists who have emphasized the radical implications of difference have been impatient with appeals to the "material." Judith Butler (1993) argues that *matter* and *material* are already discursive formations to begin with, since even the outside of discourse (i.e., the material) can only be thought by means of or through discourse. But this critique contains a misreading of the category of materialism within Marxism. In Marxism, the content of materiality is not located in a naturalistic substratum outside of analysis but rather in a particular organization of social relationships (Marx and Engels 1970). This organization is real, and given by history in a way that defies idealistic and ideological characterizations of it. To insist on the materiality of social life is to insist on the total dynamic of these relationships produced and reproduced in history. This dynamic includes discourse, not as a mere effect but as an aspect, just as it includes the physical. What distinguishes a materialist perspective from others is that it recognizes, in the different dimensions of life, not merely an indifference and chaos that only acquire sense within language but rather the registers of a fluid but immanent social *logic* that traverses them, as it does language as well. This approach is particularly relevant to the study of education since the basic meaning of schooling is expressed simultaneously as cultural practice, discursive process, and political-economic structure.

In caricaturing materialism as a mere essentialization of the "economy," narrowly conceived, common misunderstandings of this perspective ignore the way that materialist analysis *brings together* different dimensions of experience and reveals their shared participation in a historical process that cannot be identified with any single dimension. In this regard, both to push against common and narrow understandings of materialism and to clarify the political possibilities of difference, an analysis is necessary that takes up the problems of power and otherness within a focus on the embodied ecology of political life. In this regard, I take the writing of Fanon as a fundamental political analysis, one that should be located at the center of the materialist tradition. For example, Fanon's theory upsets the immaterialism of the dominant psychoanalytic paradigms in finding the sense of psychopathology at the level of the body itself: "At the extreme, I should say that the Negro, because of his body, impedes the closing of the postural schema of the white man—at the point, naturally, at which the black man makes his entry into the phenomenal world of the white man" (1967a, 160). This reframing of neurosis in terms of the corporeal—and specifically the *raced*—body is at the same time an understanding of individual suffering as essentially social. For colonized people, Fanon argues, "every abnormal manifestation . . . is the product of [their] cultural situation" (1967a, 152)—that is, of the logic of racism. The social, the psychological, and the economic as well, express a logic of domination based not only on the marginalization of different voices but also on the active organization of violence against Black bodies and souls.

Unlike the common understanding of difference as a sphere of competition for relative degrees of sanction or legitimacy, especially in education, Fanon describes difference as the index of an absolute cleavage of the social, the axis of a historical and active brutalization. In this context, existential problems are immediately determinate political ones, physical ones, and embodied ones; no longer can theoretical consciousness "flatter itself that it is something other than consciousness of existing practice," in the terms of Marx and Engels (1970, 52). In other words, even as a matter of rhetoric, difference has to be understood as a historical process, concretely and violently lived out (and through). And identity as a function of difference is recast from a structure of choices to a historical problem, inherently linked to the production not only of psyches, but also cultures and economies: "This European opulence is literally scandalous, for it has been founded on slavery, it has been nourished with the blood of slaves and it comes directly from the soil and from the subsoil of that underdeveloped world" (Fanon 1963, 96). Difference here is not an arbitrary effect of identification but the result of racism, colonialism, and capitalism as encompassing structures and processes. If oppression is a matter of material production in this way, then resistance to it must be a material production of another kind. And if domination always acts on living bodies—reducing, immobilizing, and violating them—then resistance is a matter as well of the release, as Fanon puts it, of a "muscular tension" (54). Negation is a necessary moment of liberatory movement in the context of a society in which all positivities are given as a result of exploitation.

However, there is a strong positivist streak in progressive and left-liberal approaches to difference in education, which are often organized around the idea of *success* or achievement, and an insistence on its possibility for all students. In these accounts, the caring of culturally relevant teachers for their students is expressed in a commitment to and belief in the success of all: "Rather than aiming for slight improvement or maintenance, culturally relevant teaching aims at another level—excellence—and transforms shifting responsibility into *sharing* responsibility. As they strive for excellence, such teachers function as *conductors* or *coaches*. *Conductors* believe that students are capable of excellence and they assume responsibility for ensuring that their students achieve that excellence" (Ladson-Billings 1994, 23). This is definitely an urgent emphasis against the low expectations that teachers have consistently been shown to demonstrate in relation to students of color. However, this framing of the issue risks ignoring the logics that overdetermine the very idea of excellence around which teacher expectations are supposed to be structured.

Success and *excellence* in U.S. schooling are not isolable qualities but instead are part of larger economies that encompass them. As both a concept and a process, "success" both requires and conceals the *opposite* that is part of its very structure. In systems of educational assessment, competition, and credentialing, as well as in the larger logic of social life under capital, "success" means the privilege of a triumph against that which negatively constitutes it, namely "failure." Failure is the necessary condition of success, not simply as it happens, but as the essential anatomy of the idea. Therefore, it is necessary for critical educators to recognize not only that all students are capable of achievement but also that achievement is itself constructed as the property of a particular class and color. In this sense, "excellence" is part of a discourse that systematically excludes children of color from the privileges of achievement (De Lissovoy and McLaren 2003).

The discursive economy of "success" reveals both the structure of ideology in capitalism as well as the Manichean logic of colonialism.

In capitalism, the success of a few is predicated on the immiseration of the majority, a fact that never emerges in the celebrations of those who are singled out as "excellent." At the same time, the coupling of success and failure is fundamentally tied to the logic of racism, and the material and symbolic privileges of "achievement" are awarded on this basis. The only analysis that is adequate is one that can encompass the relationship between these two systems, as Fanon's does. In the simple deploring of inequities that characterizes mainstream thinking about education and the tinkering with policies and initiatives that aim to temper the disparities around the edges, the dominant understanding dooms itself to failure as far as real transformation is concerned, since this understanding does not confront the logic that produces both the "problems" and the "solutions" at once.

Fanon reveals this contradiction in reference to education itself, a topic he does not otherwise write much about, in a passage in "Letter to a Frenchman":

> That I should say for example: there is a shortage of schools in Algeria, so that you will think: it's a shame, something has to be done about it.
> That I should say: one Arab out of three hundred is able to sign his name, so that you will think: that's too bad, it has to stop.
> Listen further:
> A school-mistress complaining to me, complaining about having to admit new Arab children to her school every year.
> A school mistress complaining that once all the Europeans were enrolled, she was obliged to give schooling to a few Arab children. (1967b, 49)

Here a reflexive condemnation of educational inequality is immediately exposed by an expression of the fundamental racism that underlies this inequality. If this racism is usually less dramatically overt in the United States, it nevertheless continues to organize, at a structural level, the system of public schooling. For change to be a possibility, it is the structure and logic of racial domination that must be exposed and analyzed rather than the mere fact of differential privilege (Leonardo 2005). The irony of efforts aimed at improving the educational resources, opportunities, and competitiveness of students of color is that the system of opportunity and competition is organized around their exclusion. In that regard, critical pedagogy has to *deconstruct* normative ideas of success and excellence at the same time as asserting that they are possible for all students.

Elite and Subaltern: Revolutionary Leadership and Educational Authority

The complexity of Fanon's analysis of oppression described above is also present in his discussion of liberatory movements, and this complexity reappears as well for critical pedagogy in its effort to imagine a truly oppositional and democratic form of teaching. Fanon shows how the contradiction of racism, in the colonial situation, appears to displace that of capital and labor as the essence of the social and to become the content, in a way, of class struggle itself. However, the contradiction between bourgeoisie and proletariat nevertheless persists in latent form and reemerges in the postcolonial period as new forms of indigenous elitism restore the urgency of the struggle against capitalism as such. In this way, Fanon describes a double-jointed structure in the movement of liberation that succeeds in organizing the people into a coherent organization and yet contains within itself the incipient contradictions of class and power between elite and subaltern that must themselves be confronted in the course of the struggle.

Being attuned to these difficulties is crucial for critical pedagogy as well. Critical pedagogy must be able to interrogate itself in a principled way with regard to the forms of leadership it proposes in classrooms, and it must also be vigilant with regard to efforts aimed at assimilating its own principles, in distorted form, into (neo)liberal currents in education (McLaren 2000). This attention to both the internal and the external contradictions of liberatory projects, and a steadfast commitment to think through them, is essential in educational movements no less than in politics generally. On the one hand it is possible, as some have suggested, that critical pedagogy may sometimes reproduce in its own practice some of the authoritarianism it aims to combat. On the other hand, it is also clear that this tradition is vulnerable to being co-opted by a range of purposes quite alien to it. An interesting case is the effort to use the tools of critical pedagogy for the education of the elite, for example in the training of future managers in MBA programs (see Currie and Knights 2003). This co-optation may represent a deep contradiction, but it also shows the originality and fluidity of capital and cautions the tradition of critical education to be vigilant against appropriations that subvert its political commitments.

The dilemmas described above have often overtaken the process of liberation—for example, when those in whose name it is undertaken are silenced and erased in the very pronouncements of its "leaders." Fanon is famous as a theorist of anticolonial struggle, but in his central work, *The Wretched of the Earth* (1963), he is equally concerned with sketching out the way this revolution has been typically betrayed by the nationalist bourgeoisie once it attains power:

> The people who for years on end have seen this leader and heard him speak, who from a distance in a kind of dream have followed his contests with the colonial power, spontaneously put their trust in this patriot. Before independence, the leader generally embodies the aspirations of the people for independence, political liberty, and national dignity. But as soon as independence is declared, far from embodying in concrete form the needs of the people in what touches bread, land, and the restoration of the country to the sacred hands of the people, the leader will reveal his inner purpose: to become the general president of that company of profiteers impatient for their returns which constitutes the national bourgeoisie. (166)

This contravention of the goals of the movement is not a random turn of events but is connected to the logic itself of the revolution, to its organization around an elite cadre who claim the authority to speak immediately for the desires of the mass. By contrast, a materialist commitment holds to the lived experience of the people against the assimilation of their aspirations to the projects of the powerful.

The dynamics of this political assimilation have been studied by historiographers of formerly colonial societies. Ranajit Guha (1988) has described how official narratives of the nationalist movement in India reproduce the elitism, at the level even of rhetorical structure, of colonial British accounts of Indian history. Guha shows that both the colonial and the nationalist accounts ignore the decisive contribution of subaltern political activity—the purposive insurgency of *the people* themselves. The lesson here is that is that any reading of social reality or project for its transformation needs to acknowledge the gaps and impasses that partly structure its own pronouncements. Such pronouncements upon and about the subjectivity of the oppressed, which construe this subjectivity as consolidated, knowable, and transparent, may conceal an effort to assimilate, through misrecognition, the experience of the other (Spivak 1988a). Similarly, Dipesh Chakrabarty (2000) has criticized the historicist underpinnings of Eurocentric liberal and left accounts of postcolonial societies, in which the latter appear as underdeveloped or deformed versions of a particular experience of European modernity.

In the context of contemporary neoliberalism, the *class* dimension of the elitism of bourgeois nationalist movements is exposed again in perhaps an even more marked fashion than in Fanon's day (Bond 2005). As the programs of once-radical organizations accommodate themselves to the demands of global capital (as in the case of the African National Congress in South Africa, or the Workers' Party in Brazil), it becomes clear that the claim to speak for the people must be critiqued for the way that it rhetorically erases the contradiction between the capitalist state and the masses. Indeed, neoliberalism's success as a class offensive has importantly come from its ability to represent itself as a progressive process of social transformation—represented in a parade of new initiatives and institutions aimed at "development" and even "empowerment."

These senses of elitism are important for liberatory projects in education to confront in their own sphere. To begin with, the idea that through the intervention of critical teachers students can be transformed from the passive objects to the active subjects of history may share a common structure with the elite narratives of nationalism and modernity that have been interrogated by the theorists mentioned here (De Lissovoy 2004). "Empowerment" itself, as it is often imagined in education, may assume the same universal and unified subject that elite accounts of national development emphasize with regard to politics. The idea that through liberatory education students discover their true selves and voices may assume an original *lack* in students that is already *dis*empowering. In this context, educators need to be wary of understanding subaltern or oppressed students transparently. They need to be wary of believing that they can know or decide who students ultimately are. Lisa Delpit (1995) has criticized progressive educators for the facile assumption that they understand, by virtue of a critical orientation, the needs of students of color. Educators involved in liberatory projects need to be sensitive to the inherent complexities that are involved in solidarizing with students without thereby giving up on radical educational praxis and movements of resistance.

On the other hand, in the neoliberal moment, there is an external crisis for critical pedagogy to the extent that its initial radical commitments have been significantly diluted as it has been incorporated into the academy and, to a lesser extent, into schools themselves. Partly, this has resulted from a watering down in mainstream contexts of the Freirean tradition as can be seen in the overemphasis on an ill-defined notion of dialogue and a purposeless imperative to "cooperation." There is also the proliferation of the ambiguous discourse of "social justice" that now organizes teacher education programs and entire colleges of education across the United States while reserving for itself a determined vagueness that often acts to obfuscate the very social forces that support the injustices that social-justice efforts presumably aim to eliminate. The language of social justice, *in practice*, frequently operates in a way that prevents a militant interrogation of power to the extent that it projects an image of progressiveness while undercutting the more charged and difficult task of naming and investigating systemic racism, for example, as a concrete problem (as opposed to "inequity"), not to mention capitalism or imperialism (as opposed to "corporatization"). This, too, is an interested misrecognition, as the generous intentions of professional intellectuals and educators are taken to be an accurate reflection of the desires, needs, and historical vocation of the people. As in the problematic traced by Fanon, an apparently unifying language of elevated and collective purpose conceals the differential location of elite and subaltern; indeed, this language can even become a tool for the consolidation of the elite purchase on power.

The Politics of Dialogue

Education scholars tend to diagram the problem in the current state of affairs in schools and society and then rather easily move to a proposal for a better situation or system. For example, basic resources and opportunities to learn are unequally distributed in schools; therefore, we should demand that more opportunities be provided to those who lack them (Oakes 2004). With regard to pedagogy itself, new teachers are exhorted to engage in critical curriculum building and to become agents of reform against the narrowing of the meaning of instruction (Cochran-Smith 1991). What is often missing is a narration of the process of liberatory practice, which is always a moving *towards* the liberatory, never the creation of a static image of it, and which is always the story at the same time of the persistent system of obstacles that this movement encounters. Without recognizing the determined systems of interests that are constitutively opposed to the equalization of conditions and opportunities, theorists unwittingly doom their proposals from the start. Educators must be able to recognize the global projects of neoliberalism, capitalism, white supremacy, and colonialism, as well as those classes and institutions that represent their interests in the arena of schooling, if we are to have a chance at overcoming the effects of these systems. Outside of a sensibility to the dynamic *economies* of both oppression and resistance, democracy becomes an ideological figment.

In his description of the stages of anticolonial movement, Fanon narrates a praxis informed by this dynamic sensibility. Fanon's praxis is a dialectical movement in that its truth does not belong to one formulation but develops through a series of positions, which as concrete engagements with an overdetermining social and psychic reality are never fully successful. Thus, to be colonized is to be forced to identify with the oppressor; to resist this, the colonized person must assert his or her absolute difference; to do so is to discover that this difference is empty and immediately recuperated in the logic of domination; against this recuperation, the self must be asserted again, not as a simple negation of the oppressor but as something new; but with what resources can that be imagined? Homi Bhabha writes that "Fanon is the purveyor of the transgressive and transitional truth. He may yearn for the total transformation of Man and Society, but he speaks most effectively from the uncertain interstices of historical change" (1994, 40). There is no solution outside of the active working through of the problem.

However, it is important to distinguish this open-ended praxis from an ethic of complete indeterminacy. In the focus on the crossing of borders and the creation of new meanings that has become important in critical pedagogy, we need to ask through what concrete *process* the educator's understanding becomes an active incitement to dialogue. How do the relationships of power and the stakes of different articulations come to be at issue in the space of teaching? In the terms of Paulo Freire, this is the problem of "codification," i.e., the production of the "*objects* which mediate the decoders in their critical analysis" (1997, 95)—in other words, how the essential contradictions of a historical situation are thematized in a way that makes them available for exploration and critique by students. In political terms, the point is that a theory of liberation, and liberatory teaching, involves not only specifying principles of democratic relationships and the forms of understanding they produce but also envisioning the program that enacts these principles. Critical pedagogy's response to the current criminalization of youth, for example, must include more than an insistence on resuscitating a culture of democratic deliberation; it should include projects that investigate and confront federal legislation and municipal ordinances, sentencing guidelines, media representations, and artifacts of popular youth culture as well—not in a reflexive gesture of protest but as a curriculum of engagement that initiates problems and possibilities rather than resolving them.

In addition, democratic pedagogy as material practice has to be able to envision the process of the teacher's own involvement as a coparticipant in the space of narration and critical self-interrogation. In Giroux's terms, "By being able to listen critically to the voices of their students, teachers also become border-crossers through their ability both to make different narratives available to themselves and to legitimate difference as a basic condition for understanding the limits of one's own knowledge" (1992, 34–35). But far from being a latent capacity that teachers simply need to make use of, this sensibility is the result of a difficult process of self-criticism and transformation that a theory of liberatory pedagogy must also be able to conceptualize. There are two important processes to consider in this regard: (1) the existential crisis teachers have to pass through in the transition from official authoritarian identifications to a revolutionary commitment; and (2) the actual dynamic movement of interruption and reframing of the teacher's voice in the classroom dialogue. In particular, recognizing teachers' *inclusion* in the discursive space of teaching suggests rethinking the shape and limits of the authority that is reserved for them even within critical approaches in pedagogy.

Fanon provides a crucial conceptualization of this process in his notion of a "fighting culture" that unifies intellectuals, militants, and the people in a common struggle for liberation. The organizational principle of this culture is a process of democratic dialogue, which importantly calls the intellectuals away from the comfort of their own sureties and into a profound (and destabilizing) dialogue with the people. This is the turning point in the oppositional movement, as Fanon describes it in *The Wretched of the Earth*, as different sectors are drawn together in an encompassing project. Hussein Adam (1999) emphasizes that, for Fanon, this is a deeply decentralized process, not directed from above, through which the slow development of the nation takes place in a genuinely consultative fashion. In Fanon's description in *A Dying Colonialism* of the importance of radio in the course of the Algerian struggle, even the interruption of revolutionary broadcasts by the French creates a space for the participation of the masses in imagining and constructing the missing voice of the movement: "A real task of reconstruction would then begin. Everyone would participate, and the battles of yesterday and the day before would be re-fought in accordance with the deep aspirations and the unshakable faith of the group. The listener would compensate for the fragmentary nature of the news by an autonomous creation of information" (1965, 86). Here the very gaps in communication create the space for the emergence of a more powerful and authentic adherence of the people to the movement. Nigel Gibson (1999) argues that Fanon intends to emphasize that the French jamming of the broadcasts disrupted the very authority of the movement, as the leaders present in any audience were forced to negotiate, along with other listeners, the meanings of what they had listened to. In contrast, perhaps, to a notion of leadership as deriving its authority from a seamless communion with the people, in this account this authority must be continually and locally renegotiated in creative fashion.

This is an important lesson for critical educators to learn, or relearn. It is possible that conceptions of democracy that emerge from the educational context may have a tendency toward authoritarianism, given the structures of institutional power that usually set the stage for the encounter between teachers and students in schools. And in the present context, in which teachers are instruments of a hyperauthoritarianism at the level even of the fundamental organization of cognition (through micromanaged and prepackaged curriculum and assessment), it may be important for conceptions of democratic educational engagement to start from models outside of formal instruction. It is crucial for critical pedagogy to always remember its affiliation to the larger movement and to the moment of decentered and collective struggle, and to internalize this struggle as the controlling metaphor of its own approach to dialogue and solidarity in teaching situations. Fanon's emphasis on the complementary roles of participants in the movement, including peasants, intellectuals, and workers, recalls Rosa Luxemburg's (2004) insistence that the only authority that really counts, in the process of learning through struggle, is the "mass ego" of the working class (as opposed to the authority of the professional cadres), which can only authentically learn through its own mistakes and experiments.

Furthermore, it might be salutary for critical educators, most of whom have been shaped by the experience of a wielding an institutionally guaranteed power in the space of dialogue (that of the teacher), to remind themselves that teaching is only one moment of potential struggle and that the position of the critical teacher is the position of only one kind of participant in it. In this light, the point of critical pedagogy is less for the teacher (as leader) to steer the students' learning toward the proper (critical) analysis and more for both students and teachers to be provoked to kinds of investigation that can increase the knowledge and possibilities for collective movement, in which both are participants and leaders in a democratic "life in common" (Hardt and Negri 2004). For example, youth-led movements against militarization of schools and the impoverishment of the curriculum by standardized testing regimes are opportunities for engagement and learning in which teachers can act as crucial resources without occupying the position of leadership. At the purely methodological level, this suggests the importance for critical pedagogy of rediscovering the radical potential in the decentralized and Deweyan notion of curriculum as project and activity—reconceptualized in this case, however, as *revolutionary* project and activity within the context of a wider movement.

Conclusion: Materialism as Humanism

The vulgar version of materialism is the idea that there is no truth except in the cold, hard reality of things themselves. This simple objectivism, however, is not what Marxism means by materialism. In materialism, the relevant opposition is not between ideas and things but rather between ideology and the social relationships from which it derives. The mistake of bourgeois philosophy is to suppose that truth exists prior to and apart from human history. Materialism is fundamentally an intellectual and ethical responsibility to history and coincides with a humanism that registers the suffering that has marked this history. In this tradition, Fanon condemns the violent and distorted idealism of Europe: "Leave this Europe where they are never done talking of Man, yet murder men everywhere they find them, at the corner of every one of their own streets, in all the corners of the globe. For centuries they have stifled almost the whole of humanity in the name of a so-called spiritual experience" (1963, 311).

An authentic humanism must see past these myths to the historical violence they have concealed and must materially intervene to disrupt it. The liberatory philosophy that emerges in this project is necessarily grounded in this concrete imperative: "It is simply a very concrete question of not dragging men toward mutilation, of not imposing upon the brain rhythms which very quickly obliterate it and wreck it" (1963, 314). Fanon's challenge presses European left traditions to recognize their own myopias, which have often pressed the violence of colonialism, slavery, and racism into the background. The prioritization of these histories, and of the lessons they offer, represents a tremendous expansion of the possibilities of both materialism and humanism, since attention to them reveals vastly more of the concrete logic that has characterized the historical exploitation of human beings and since the overcoming of these forms of violence opens new possibilities for what human being and human solidarity might be.

As Fanon shows, humanism is a political project that works through the materiality of social relationships and revolutionary aspiration. It is an extrapolation of *humanity* from a historical kindredness that is at the same time both difference and connection. The coherence of liberation, in this perspective, comes not from a scientific glimpse of the continuous essence of oppression but rather from the combination of struggles that find in their commonalities and simultaneity the possibility of a shared project and the outlines of a shared opponent. This historical solidarity returns to the category of *class* its original political character, indeed its human, practical, and processual aspect (Przeworski

1986). For educators, a commitment to this humanism means a continuous process, in collaboration with students, of making connections: between the racism of reductionistic, test-based "accountability" initiatives and the racism of a militarized society that uses youth of color and poor whites as cannon fodder; between the bottom-line logic of global neoliberal structural adjustment and the gutting of a stripped-down public education system by a vicious "compassionate" conservatism; between the demonization of brown people in distant "rogue" states and the vituperation against immigrants in the U.S. heartland. In studying their own and others' experiences, students and teachers can trace for themselves the shape of Fanon's "stretched" dialectic and the complicated imbrication of oppressions that it exposes in the contemporary context.

This chapter has described how Fanon's concrete analysis of the forms of oppression in the African colonial context enlarges the meanings of liberatory praxis and dialectical analysis in critical pedagogy and beyond. While Fanon's work has been indirectly important to critical pedagogy in its influence on Freire and other theorists, I believe that it should occupy a central position in this tradition's sense of its guiding exponents and its historical forebears. Indeed, without reference to the arguments and analyses of Fanon, I believe it will be difficult for critical educators to resolve the tremendous dilemmas that neoliberalism, racism, and globalization pose for education—not only taken separately (if this is possible), but more importantly in their complex imbrications and interrelationships. Fanon provides indispensable insights into the complexities of liberatory struggle and a profoundly dialectical analysis of the stages and mutations that such struggle must recognize and traverse. For educators, oppositional movement coincides with the building of solidarity in teaching, and in this regard Fanon's exposition of the principles of revolutionary-democratic dialogue is especially relevant. In the context of education as well as social life generally, Fanon's work doubles and deepens materialist analysis and unfolds new dimensions of a more complete and expansive humanism, the vision of which we cannot afford to do without.

CHAPTER 3

Conceptualizing Oppression in Educational Theory
Toward a Compound Standpoint

For educators, theoretical problems regarding the nature of oppression are reflected in important practical questions. Is the key struggle with regard to schooling basically about curriculum—is it a matter of working against sexist and racist representations, for example, in the content that is presented as legitimate and the mastering of which is the criterion for academic success? Or are the possibilities for liberatory education more or less determined by the structural support, in terms of facilities and resources, that are disproportionately available to students from wealthy families? Are such formulations themselves narrowly reformist—and is the essential task to link, in the minds of teachers and students, their struggles in schools to struggles outside, to larger movements of resistance against global processes of domination? How important in all of this is white supremacy, as opposed to poverty, or class oppression, or the gendered discourses that make schools sites of danger for lesbian and gay students and help to inculcate the violent rituals of masculinism? On the other hand, is it necessary for teachers first of all to engage in the work of deconstructing their own official interpellation as professionals, which already flattens them spiritually, and destroys the possibility of creativity in oppositional teaching? Or can the problem of education really only be grasped in a larger frame, as an ideological apparatus comprised of the media, popular culture, churches, and so forth, in addition to schooling?

Being able to respond to these questions depends on having an understanding of the sociopolitical context of education and in particular how oppression constructs this context. In this chapter, I specify the general theoretical problems raised in the preceding chapters through an examination of several different conceptions of the nature of oppression that are important in contemporary critical approaches to education, as well as in progressive and left scholarship more generally. While these are not the only important conceptions, they can, in their differences, usefully organize the frame of a discussion of oppression. I give an account of each, and of some of the problems that each motivates, as well as considering the important commonalities and tensions between them. This investigation then suggests, in pulling both from what is present and what is absent in the accounts I have examined, a new theoretical position on the question of oppression in education. In this effort, I seek to respond to recent calls for a rejuvenation of critical approaches in education and a creative rethinking of their possibilities. My analysis and proposal in this chapter demonstrate a way forward for several urgent aspects of this project: the need to challenge the finality of given approaches and their compulsive repetition (Kumashiro 2002), the importance of linking diverse critical discourses in education in a process of mutual interrogation and cross-fertilization (Leonardo 2002), and the necessity of reconnecting theory to contemporary educational struggles and oppositional movements (Leistyna 2004). The framework that I articulate responds to these imperatives and links them organically in the context of an underlying commitment to educational research oriented to social and political transformation.

The paradigms of oppression in educational research that I consider here are: (1) the model of *cultural hegemony*, important in the tradition of critical multiculturalism; (2) the analysis of oppression as a matter of *capitalist accumulation*, which characterizes Marxist approaches; (3) the model of oppression as a matter of regulatory discursive norms (i.e., oppression as *discursive effect*), associated with poststructuralism. These three approaches to the problem of oppression in education are very different, and I explore the productive tensions between them by means of a comparative analysis of the work of three central scholars associated with these models: Sonia Nieto, Peter McLaren, and Valerie Walkerdine, respectively. Although I discuss the work of individuals, my purpose is to motivate a conversation among approaches rather than between personalities, and to foreground research streams that are broadly shared and influential beyond the work of these particular scholars. I am sympathetic to all of these approaches and share the goal of social transformation that is important in each. The contradictions between them correspond to practical dilemmas in efforts to articulate forms of liberatory education. However, underneath what sometimes appear to be their incommensurable logics there is an important commonality. Each of these accounts posits a singular principle as the truth of oppression. This principle, in each case respectively, in accounting for the world that the text elaborates, limits the extent of that world arbitrarily. The meanings that a combination of such worlds might suggest are thereby foreclosed. I argue that a *compound standpoint*—a perspective that resists the prioritization of singular principles of the social in favor of a focus on their simultaneity and mutual imbrication within historical processes of domination—opens up a universe of greater possibility to radical theory and praxis.

Standpoint Theory and an Expanded Materialism

Critical and progressive pedagogical paradigms approach the problem of oppression and education from a variety of directions whose fundamental bases often remain unexplicated, leading to arguments that usually generate more heat than light. I suggest that we can understand these different paradigms as deeply connected to *standpoints*—epistemological-political formations situated in specific social experiences rather than simply worldviews—and that in this way we can better appreciate the internal validity of each perspective as well as its limits. Furthermore, we can then set about building a more comprehensive political-educational philosophy that can combine the analytical strength of each account in a compound perspective that starts from a recognition of the actual interpenetration of different forms of oppression in historical passages of domination. This enlarged sense of standpoint can be linked to programs for liberatory educational praxis that are themselves both theories of knowledge (in terms of how students and teachers come to critically understand their world and what the nature of such understanding is) and methodologies of effective educational practice.

Standpoint theory has its roots in the work of the Marxist philosopher Georg Lukács (1971). For Lukács, proletarian or revolutionary consciousness, in its moment of truth, is a universal consciousness, since it cognizes and transforms reality in a way that emancipates humanity as a whole. In his account, the leading edge of revolution resides less in the working class as an objective social formation than in a proper understanding of society; such an understanding should point the way toward an overcoming of the logical partiality of bourgeois thought and capitalist relations of production. Authentic liberation is a matter of the proper standpoint on the whole, a standpoint that resists its fragmentation and reification into falsely separate parts and identities. The authentic apprehension of the totality is identified with the overcoming of capitalism and class society, as revolutionary consciousness and action coincide in a complete praxis.

Lukács' analysis has been appropriated and reinvented by feminists in ways that have important results for educational theory. Hartsock (1983) transposes the notion of class position in Marxism onto the category of gender in arguing for a *feminist* political and epistemological standpoint. As with the proletarian consciousness described by Lukács, a feminist standpoint does not arise automatically but instead must be struggled for, and achieved, against dominant ideologies and social practices. Thus, a feminist standpoint is an "interested position" (Hartsock 1983, 285) that grows from and reflects upon the material activity and experiences of women (though it is not identical with those experiences). There is no simple equivalence (in terms of value or objectivity) between a feminist *standpoint* and a masculinist *worldview*, since the former arises from the activity, reflection, and struggle of an oppressed group while the latter represents the dominant view that abstracts from and obscures these realities. To the extent that a feminist standpoint articulates fundamental truths about society that are systematically excluded and obscured by dominant discourses, it allows access to something like the essence in this duality, the masculinist vision representing the mere surface or appearance of social truth. Similarly, critical educational paradigms strive to illuminate systematically hidden aspects of schooling systems, as well as of the ideologies and discourses that support and circulate in them. Recognition of these repressed contents (e.g., white supremacy or commodification) does not simply add to our understanding of educational processes; it represents a fundamental point of departure and foundation for a critique of oppressive practices.

The original understanding of a revolutionary standpoint (whether in the Marxist or feminist conception) yields to its own inevitable complexities. While Hartsock (1983) makes the case for sexual difference and oppression as fundamental in organizing social life, once the distinct social positioning of women can be seen to allow for the development of a specifically feminist standpoint, there is no reason why other social locations cannot be recognized as capable of producing their own epistemologies and political philosophies. As Harding (1993) points out, once we see that there is no universal man, we must see that women too are never homogeneous but always differ in terms of class, race, and other factors. Each distinct social position, to the extent that it is involved both in different material activities and different opportunities for struggle, can (potentially) be productive of a standpoint. And in fact, feminist theorists have proposed standpoints that respond to the experiences of specific groups of women more usefully than an undifferentiated (and "universal") feminist standpoint. Collins (2000), for example, has proposed an epistemology of Black feminist thought proceeding from the unique experiences of Black women. Likewise, educational theories grounded in the histories and struggles of different groups of students can differently afford and constrain possibilities for understanding and action. A refusal to pronounce upon the nature of truth for everyone is implicit in this conception. However, where there is difference there is also the possibility of combination: bringing together different standpoints can potentially increase their power as well as allowing for the possibility of reciprocal critique, as I describe in this chapter.

Traditional standpoint theory confronts certain intractable obstacles, however. Since a standpoint is thought of as emerging from the experience of a particular social group, each standpoint is ultimately inadequate for understanding social life outside of the characteristic experience of that group. In addition, the assumed constituency of a standpoint can seem to be falsely universalized (e.g., in assuming "women" as an undifferentiated and unproblematic category). Also, in its focus on consciousness and knowledge as the starting point for politics, standpoint theory risks obscuring the necessary materiality both of domination and liberation as concrete historical processes. For these reasons, in order to bring together the insights of different oppositional paradigms toward a synthetic or "compound" standpoint, an expanded materialist grounding is necessary that attends to the complex historical determination of oppression in concrete processes of conquest and domination. The work of anticolonial theorist Frantz Fanon (1963, 1965, 1967a, 1967b) provides crucial resources in this regard, since his analysis demonstrates the way that economic, political, cultural, and psychical registers are complexly interwoven and even identified in the historical problematic of colonialism. Like standpoint theory, Fanon's work emphasizes the importance of consciousness; however, the conception of decolonization that Fanon proposes never separates consciousness from the dynamic historical context in which it finds its meaning. This expanded materialist approach makes it possible to step out of the apparently mutually exclusive epistemological worlds of different standpoints and to combine their most useful insights with reference to a common, if complex, historical process of conquest and struggle against it. This perspective is especially useful for understanding education, in which economic, cultural, and ideological forces interpenetrate and determine each other.

The framework outlined above orients the following analysis in several ways. It makes it possible to understand educational paradigms as politically and socially situated, with in each case characteristic possibilities and constraints that are not incidental but rather intrinsically related to the epistemological ground of each approach. In addition, however, it underlines the importance of attending to the common root of different forms of oppression, and thus of the liberatory standpoints that seek to counteract them, in historical passages of conquest and domination. Finally, this framework is the basis for the expanded analysis of oppression in education that I offer in the discussion section, which undertakes an exploration of the possibilities for multiplying the power of individual paradigms through the notion of a compound standpoint.

Oppression as Cultural Hegemony

The first paradigm of oppression in educational theory that I consider here is that which sees it as a matter of cultural hegemony. The cultural hegemony model has developed in response both to popular conceptions of racial and cultural prejudice as accidental individual pathologies, as well as to overly psychologistic treatments of racism in educational research. It places on a firmer theoretical footing the analyses of the radical democratic and civil rights traditions that have protested pervasive social inequalities, as well as offering an alternative to more class-based approaches on the left. This model is prominent in critical multicultural education in particular, as well as in sociocultural approaches to classroom processes. The central conception within this perspective, and in the work of Sonia Nieto in particular (e.g., Nieto 1998, 1999, 2002, 2004), in regard to the framing of the nature of oppression in its various manifestations as racism, sexism, linguicism, et cetera, is the idea of a dominative power acting consistently and pervasively in its own interest, i.e., a European American cultural hegemony. Nieto's naming of white supremacy as an omnipresent force in schools insists that racism needs to be understood beyond the level of individual prejudice: "Prejudice and discrimination, then, are not just personality traits or psychological phenomena; they are also a manifestation of economic, political, and social power. The institutional definition of racism is not always easy to accept because it goes against deeply held notions of equality and justice in our nation. . . . Racism as an institutional system implies that some people and groups benefit and others lose. Whites, whether they want to or not, benefit in a racist society" (2004, 38).

In schools, this systemic racism results in a series of structural and pedagogical injustices, including tracking, retention, low expectations for students of color, and generally the privileging of the needs of white students. Similarly, Nieto describes culture as always more than an individual or even group matter. Culture is always involved with power, and specifically with a system in which those in power can construct dominant values as "normal" while others are forced to undergo a process of "deculturalization" in the context of a "failure to acknowledge the important role that culture may have in students' values and behavior, and consequently in their learning" (1999, 34). This hegemony is reflected in deeply biased curricula and practices in schools, discontinuity between home and school contexts for oppressed groups, and the pathologization of different approaches to learning.

The validity of the systemic analysis in the model of cultural hegemony is regularly confirmed by evidence that shows that disparities in educational opportunity are widespread and profound. For example, based on reports from teachers, schools in California with the highest concentrations of African American and Latino and Latina students are much more likely to have uncredentialed teachers, poor working conditions, and inadequate and unsanitary facilities (Harris 2004). Similar problems are to be found throughout the United States. In addition to structural inequities, Nieto confronts the less quantifiable ideological organization of curriculum and pedagogy in favor of dominant groups and undertakes a principled identification of the source of these injustices, naming racism as a system in a way that mainstream educational research often neglects to do. In the case of language diversity, for example, she emphasizes the pervasive lack of primary-language instruction available to English language learners, as well as the lack of specialized training for educators working with these students. This institutional lack of support is connected to widespread negative attitudes toward the maintenance of primary languages for immigrant groups. These phenomena, Nieto argues, are not merely incidental; underlying them is "the relative power or lack of power of various groups in our society" (2004, 226). This account thus offers a pointed political analysis of the connection between cultural violence and structural disparities.

However, there are important difficulties associated with the conception of power that is at work in the model of oppression as cultural hegemony. Are different forms of discrimination and domination all susceptible to being understood in terms of culture, as the rubric of multi*culturalism* implies? The problem is not so much that the specificity of different oppressions may be overlooked but rather that quite distinct kinds of sociopolitical processes may be lumped together erroneously. This is the case with the effort to assimilate social class to the analysis described above. In this way, "classism" becomes another example of the systematic privileging of one social group, this time on the basis of socio-economic status, that is familiar in racism—i.e., "discriminatory beliefs and behaviors based on differences in social class; generally directed against those from poor and/or working-class backgrounds" (2004, 435), as the glossary in Nieto's key text *Affirming Diversity* defines it. However, the problem with class is not mainly that certain classes are discriminated against but rather that different classes exist at all. There can be arguments about class cultures, and the kinds of distinct values and resources they create, but it is not a radical observation that the actual production of subordinate and superordinate classes per se is in no way to be celebrated, as the protest against a particular class *bias* seems to suggest.

The organization of the analysis in this model around the category of culture also results in a less-than-complete account of gender in education. Gender is important not only in terms of the differential access of female and male students to educational opportunity; schools are important sites as well for the interpellation itself of gendered subjects. Furthermore, gender interacts with race, culture, and class in complex ways, so that to consider these factors apart from gender may be to fail to give a satisfactory account of these processes even in themselves. How can seemingly antiracist politics or programs reproduce sexism (Castillo 1994)? How do girls and boys learn very different and "gender-coded" strategies for negotiating the divide between working-class home culture and bourgeois school culture (Arnot 1982)? How is the patriarchal organization of legitimate and illegitimate identifications around gender, and the construction of queer identities as outlaw and "abject" (Butler 1993), connected to the alienation of all students in school?

There are also important problems with the conception of racism itself in this paradigm. In the model of cultural hegemony, there is a move away from the notion of racism and cultural bias as mere pathologies. Recognizing its participation in a form of political rule, this paradigm moves toward the idea of racism as the organized and willed "project" (Omi and Winant 1994) of a group or society—a project marked by its own reason and purpose. Nieto argues that schools are "organized to reflect and support the cultural capital of privileged social and cultural groups" (1999, 55). In this context, "the confirmation of the dominant culture's supremacy represents a *symbolic violence* against groups that are devalued" (2004, 312) and thus a validation of the rule of the dominant. Nieto's framing of racial and cultural oppression as purposive is a deliberate challenge to those who characterize these phenomena as simply irrational, isolated, or deviant.

However, the "project" of racism consists of more than symbolic violence and the abstract consolidation of power. Any satisfactory account of racism has to include a reckoning of the material logic of plunder and exploitation. This does not mean that racism is just a simple tool in the service of accumulation—or even that its production has to be finally understood in terms of the development of capitalism, as the effect of the exploitation of non-European peoples by the emerging bourgeoisie—as Oliver Cromwell Cox (2000) and other Marxian scholars have argued. Nevertheless, liberatory and antiracist education has to include these dynamics in its definition of racism, as well as starting from the fact of historical conquest.

In the United States, race and culture are thought to inhere as forms of social or psychological identification in persons or communities. What is erased in this framing, however, is the deep interrelationship between race and the categories of land, territory, and nation. It is important to recognize the way that race and culture extend beyond identity to encompass the *territorialized* struggles of collective political subjects. By contrast, Nieto's own analysis of the legacy of colonialism for students from oppressed groups (e.g., Nieto 1998) foregrounds the symbolic dimension and the register of representation. Even the formula of institutional discrimination, which is at the heart of Nieto's conception of racial and cultural oppression, is inadequate to describe the history of conquest that continues up to the present for indigenous peoples, as well as the systematic violations of native sovereignty, in both material and discursive terms (Lyons 2005). The trope of *inclusion*, which in more or less sophisticated forms animates the various discourses of multiculturalism, has been, as annexation and assimilation, the historical mode of violence and appropriation enacted against Native Americans (Grande 2000). What is the relationship of the various stages of multicultural education that Nieto sketches—"tolerance; acceptance; respect; affirmation, solidarity, and critique" (2004, 384)—to the necessity for resistance and *autonomy* as strategies and values for indigenous peoples in particular and for other oppressed groups? In addition, in what ways do contemporary expressions of racism, at both a discursive and material level, represent new and organized offensives to subjugate, control, and expropriate? How does the representation of Latino and Latina and African American youth as violent and monstrous, a representation enforced in urban schools through enormous security apparatuses and draconian disciplinary procedures, provide in symbolic terms the resources for U.S. whites to imagine themselves as (inversely) both right and good, according to a dialectic similar in many ways to the colonial one described by Fanon (1963)—and thus the sanction for the propagation of a racist imperialism both domestically and globally?

Oppression as Capitalist Accumulation

The view that capitalist accumulation, as a system of political economy, is at the root and constitutes the essence of political and social oppression, including in education, has very much informed the tradition of critical pedagogy and more radical approaches to teaching generally. Recently there has been a renewed interest in a Marxist perspective grounded in political economy within cultural studies generally; in the field of education this is most importantly the case in regard to the work of Peter McLaren (e.g., McLaren 2000, 2001, 2005; McLaren and Farahmandpur 2005). This perspective differs from the cultural hegemony model I have just discussed in its explicit attention to economic questions and by its materialist texture generally, even in relation to pedagogy. Whereas the cultural hegemony model deliberately foregrounds racism, as well as structural and ideological processes of cultural discrimination, the Marxist view focuses on the capitalist imperative of accumulation of surplus value, which drives imperialist interventions and neoliberal austerity regimes abroad at the same time that it promotes privatization in the United States along with a valorization of the ideology of individualism and consumerism in schools and in society: "Neo-liberalism has become the lodestar for the new world order's conquest of the next millennium. Wages have been compressed worldwide, income is steadily being transferred from labor to capital, corporations are struggling for a comparative advantage in cheap labor and in acquiring state-subsidized access to national resources while abandoning public obligations to the poor. The result is class polarization, downward mobility, and class secession" (McLaren 2000, 20).

McLaren describes this accumulative frenzy as having reached a fever pitch due to a crisis in profitability brought on by overcapacity. Furthermore, following William I. Robinson (1996), he argues that the aggressive penetration by capital of global society, and the increasing integration of global elites, has given rise to a transnationalism that subordinates national interests to those of multinational corporations. Social life everywhere is increasingly subject to the demands of capital, and education is a crucial arena in which this can be observed.

In this account, while capitalism possesses its own foreshortened rationality (the logic of the market and exchange so touted by conservatives), on the grand scale it is a mistake to expect it to operate in a way that makes any human sense. As McLaren puts it, capitalism "works from a discourse of metaparanoia, in which the outside world is constructed in the service of its own agenda of accumulation, of profit-making, of reproducing its own advantage and control over the marketplace" (2000, 31). This perspective allows us to contextualize contemporary conservative reforms in schooling within a global context. In this light, current initiatives to privatize schools, force the adoption of packaged curricula, attack tenure and other protections for teachers, and replace instruction with incessant and expensive standardized testing programs can all be understood as a kind of domestic "structural adjustment" of the public space of education. These policies can be seen as fitting comfortably within the larger frame of the worldwide dismantling of welfare and social security systems, the denationalizing of public industries, and assaults on workers' rights. Within pedagogy itself, the instrumentalization of teaching as pure procedure and the ideological reconfiguration of students as consumers of a homogeneous knowledge-product, can be traced to the aggressive colonization of the space of learning by the logic of capital. From the perspective of this global accumulative drive, every public space or moment is wasted to the extent that its intrinsic value does not yield a dividend, as profit, to capital.

The model of oppression as capitalist accumulation seeks to recover a historical essence from among the disparate and confounding surfaces of experience that students confront. Revolutionary education, McLaren argues, must contextualize events and discourses in the larger framework of capitalist social relations; this means engaging in a "pedagogy of demystification centering around a semiotics of recognition, where dominant sign systems are recognized and denaturalized, where common sense is historicized, and where signification is understood as a political practice that refracts rather than reflects reality"

(2005, 59). What is at issue here is the production of an oppositional analysis that sees past the ideological naturalizations of the present to its fundamental conditions. This is a pedagogy that views social processes, cultural events, and even personalities in relation to the overall organization of the relations of production of society. Teaching, then, while it seeks to create sites in which the identities of educators can be connected to the problematic of social justice, must do so against a very particular backdrop—the history of class struggle, as well as the system of capitalist relations of production. Critical pedagogy, while it uncovers a new space of imagination and action in the world, has a responsibility to make this determined historical situation transparent.

However, we might ask what the relationship of this revolutionary knowledge is to the various "common senses" of oppressed communities. In the dialectics of liberation described by both Gramsci (1971) and Freire (1997), popular understandings are not simply reinterpreted but rather negotiated in a collective process of reflection and dialogue that leads toward a transformational analysis that cannot be specified beforehand by teacher or leader. An overly rigid historical materialism, as educational program, risks foreclosing the contribution to and *crucial determination of* authentic oppositional movements by the knowledges offered by students and others who have experienced oppression. The issue here is not just that more heads are better than one when it comes to figuring out how to understand political reality. Rather, the point is that a true, or truer, knowledge of oppression will need to draw from other epistemological resources in addition to the particular logic of historical materialism.[1] What other kinds of vision and analysis have enabled radical struggle in education and elsewhere? Furthermore, this oppositional understanding may be based on something larger than a *logic*; it may come from, in addition, an image, experience, story, or feeling. How can we consider emotion, for example, as a form of knowing (Jaggar 1989), or resistant popular culture as crucial political information (Kelley 1997)? The *meaning* of oppression may be something more than its causes, genealogy, or structure—and the meaning of oppression has also to be confronted in struggling against it.

In concrete terms, how can we account for the most powerful instances of resistance in recent decades in the United States—in schools and elsewhere—and not just account for them, but hear the kinds of understanding of oppression that they have offered? How can we imagine a Marxist reading that would emphasize the priority of the analysis of many of the actual participants in civil rights, desegregation, and antiracist movements and embrace the essential theoretical resources offered by the struggle itself—namely, the exposure and repudiation in racist late capitalism of the fundamental spiritual evil that is one of its animating energies? Likewise, how can this paradigm learn, from the struggles against sexism that successive waves of feminism have waged, about the kind of political determination that patriarchy is as such? To recognize these political dimensions is not to deny that they are also the expression of class antagonisms, complexly mediated though the prisms of gender, race, and religion. But the very richness even of a class analysis of this history depends upon its responsiveness to these and other autonomous factors.

Furthermore, the absolute centering of political economy begs the question precisely of a dialectical understanding of society—in particular, the question of the relationship between surface and essence. In arguing that much recent theory runs "the risk of helping the capitalist class manage the ongoing crisis of the humanist subject rather than confronting the universalizing effects of finance capital" (2001, 704) and that it ignores the reality that working people globally share a common subjection to capitalist exploitation, McLaren seeks to return our attention to ostensibly the most fundamental level of social oppression. But when the confusing problems of contemporary culture and politics are brought back to the matrix of political economy, what becomes of the truth that these issues represented in themselves? It is not really a question of deciding where the crux of the matter is (the economy, patriarchy, or other systems) but rather of considering, in any formulation, what relationship there is between surface and ground—how exactly the apparently "secondary" is its own kind of essence and necessity.[2] The effort to pack all the various problems of the social into one of its levels (e.g., the economic) denies the very dialectical structure of this social totality—the complex *relationship* of parts that makes up the whole. These separate levels (politics, culture, ideology, etc.) will not stay put—they will inevitably float out and up again to indicate, by their very separateness, the system of relations that makes their existence, after all, dialectical. What is urgent to uncover is the relationship between the political reconfigurations of schooling (e.g., in terms of literacy instruction or disciplinary regimes) and the economic conditions that both impel and are determined by these reconfigurations.

This point can be made another way. In capitalism, the objective conditions of labor take on an alien subjectivity in relation to the worker—even a kind of personality, in the form of the capitalist (Marx 1973). So political analysis has to consider the production of class subjects and investigate the forms of relationship of these subjects. In other words, classes are important historical actors by virtue of the relations of alienation and antagonism that pose them as subjects. In this way, the economic problem is itself essentially a problem of social relationships, and the domains of politics and culture become central as social terrains in which the form of these relationships is struggled over. To insist that nevertheless struggles around cultural questions still represent in some sense the playing out of class antagonisms is very different from saying that these struggles necessarily refer back, as if to their essence, to economic processes.

Oppression as Discursive Effect

The model of oppression as discursive effect is important in the work of scholars in education and other fields who have considered how the meanings and practices in which we are able to come to be and to do are profoundly regulated by systems of norms, as discourse; these norms precede and permit the intelligibility of social subjects. In such analyses, often identified with the philosophical movement of poststructuralism, questions about subjectivity and oppression are necessarily complicated—complicated in fact by the questions themselves, which cannot be clearly separated from the truth they would get at. In regard to education, this view challenges simple sociological models that pretend to reveal transparently objective processes of socialization as the inner truth of schooling; this view also challenges progressive approaches to teaching that claim an untroubled access to the essence of students as children or learners, even for seemingly beneficial ends. One of the most powerful accounts of schooling from this perspective is provided by Valerie Walkerdine, who has explored popular culture and pedagogies of literacy and numeracy as discursive regimes that are deeply gendered, no less so (and perhaps especially) in reforms aimed at improving achievement or against bias (e.g., Walkerdine 1990, 1997, 1998). Walkerdine's work is part of a tradition of research that combines the resources of feminism and poststructuralism toward an analysis of patriarchy not simply as a mode of discrimination but as deeply implicated in the humanistic reason that has produced modern science and education in "Western" culture.

Walkerdine investigates how a series of constitutive oppositions organizes learning, and how gender is related to these oppositions. In the context of the classrooms that she has studied, educators consider "real understanding" to be the result of "natural" and unforced development and identify it with boys. On the other hand, "rule-following" or rote learning, the product of an *unnatural* diligence, is thought to be an inferior kind of learning and is associated with girls. These interpretations are motivated less by a bias in favor of boys than by a basic need to prove that girls are inferior: "We have suggested that within current school Mathematics . . . 'Woman' becomes the repository of all the dangers displaced from the child, itself 'father' to the man. The need to prove girls' mathematical inferiority is not motivated by a certainty, but by a terror of loss. In this story these fantasies, fears and desires become the forces that produce the actual effectiveness of the construction of fact and of current discursive practices" (1998, 37). This terror is located not simply in men but beyond both men and women in the discourse that regulates the way that their identities and actions are allowed to signify. Most importantly for the school context, Walkerdine describes how girls, considered as "children," are expected to learn according a pattern of "natural" development that is in the same moment forbidden to them as "women." In this way, she finds that progressive and child-centered pedagogy reconfigures the truth of learning from external performance to the internal and elusive movement of "real" understanding and so constructs learning as a matter (following Foucault 1977) of the *soul*—a soul that, Walkerdine shows, is understood at the same time to be male.

Walkerdine's account dramatically reconfigures the conundrum for progressives of deciding the priority of structure and agency in educational processes. While theories of social reproduction (for example, those of Althusser 1971 or Bowles and Gintis 1976) powerfully indicted educational systems for their complicity in enforcing social inequalities, it seemed at the same time that the power of individuals and oppressed groups to intervene was discounted (Giroux 2001). Anxieties about how best to balance this equation have resulted in what is now a kind of stale truce, in which descriptions of the pervasive force of systems of socialization are immediately countered with a symmetrical emphasis on their indeterminacy and the ability of individuals to contest them. In Walkerdine's narrative, by contrast, structure and agent are not independent or interdependent but rather indistinguishable. The kinds of effects that critics normally associate with the system are already present in the individual—and not in a way that constrains the true kernel of her self but rather as the arrangement of commandments in relation to which any possible identity has to emerge. For example, there is a special pressure on girls to behave well and follow rules. On the other hand, the discourse of progressive education constructs "natural" learning for the child as involving a willingness to break the rules; this confidence and freedom is the necessary sign of "authentic" development. In this way, "girls come to desire in themselves qualities that appear opposite to those of 'the child' whom the pedagogy is set up to produce" (Walkerdine, 1990, 76). The impossible conflict in which they are then placed is not just a matter of external biases but rather of how they can manage to successfully perform basic identifications, in the first place, as good girls and as normal children.

This deconstruction of the opposition between structure and agency in part usefully complexifies the political terrain. However, it is important to consider whether the variously deconstructive, genealogical, and psychoanalytic methodologies of poststructuralist critiques reproduce, in their own way, the immobilizing force of some of the structuralisms they aim to "trouble." If it is important to question the demoralizing effect of mechanical accounts of social reproduction, it is also important to consider the political effect of poststructuralist analyses. And while the optimism of overly voluntaristic theories of agency in education may ignore the complexity of the binds in which students are placed, what does an account that seems to undo even the collective hope and faith in transformation offer in their place—or is the offering of any specific sense of "possibility" automatically a reinscription of the inescapable grammar of power as humanistic and regulatory discourse?

Furthermore, the philosophical model that views oppression as a matter of discursive constructions has tended to problematize common-sense understandings of political struggle and opposition. This raises the question of the relationship of intellectual work to movement politics and to the understandings of those active in practical struggles. What do radical teachers and activists do with texts that often undo the political schemas that underlie their work? What is the responsibility of intellectuals in this regard? Poststructural approaches reveal the complexity and difficulty of resistance, and one obvious response is that it is necessary for teachers to be wary of apparently progressive interventions that covertly reproduce essentialist representations of oppressed communities (Kaomea 2003) and for activists to work through these questions toward forms of praxis that are more self-critical. However, it is also important to question the way that poststructural analyses tend to privilege the academic and theoretical avant-garde (De Lissovoy 2004). As they compulsively problematize popular oppositional understandings, poststructuralists often then reterritorialize left politics on purely textual grounds. As far as academics are concerned, this move undermines the possibility of the "critical organic catalyst" (West 1993b) who stands between popular movements and scholarly work and who recognizes the contributions of each sphere to discovering political truth and strategy.

The central issue here is the way that discourse-oriented approaches are fundamentally irritated by the conceptual binaries that mobilize many oppositional movements. Walkerdine calls into question simple antisexist or antibias curricula to the extent that they reify "girls" and "boys" as unproblematic and essentially distinguishable categories and because they often reproduce the idea of a lack associated with girls (even if this lack is taken to be the result of socialization). Rather than pregiven and "real" identities, Walkerdine argues, the various possibilities of "girl" or "woman" are given by their differential relationship to other signs (e.g., "child" or "teacher") within discursive chains of signification: "We are not setting out to demonstrate the *reality* or *prove* that girls *really can* do Maths or boys actually *do not* have real understanding. Rather, we are interested in how those categories are produced as signs and how they 'catch up' the subjects, position them and so create a truth" (1998, 39). But to what extent does this emphasis on the different kinds of power afforded by different positions within discourse obscure the necessity of an analysis of *interest*? In this regard, the missing discussion of race and racism in Walkerdine's analysis is symptomatic. Considering the meaning of schooling as "racial text" might challenge the coherence of Walkerdine's theoretical frame. While "race" and "culture" are themselves effects of complex and ambivalent discourses (Bhabha 1994), they are also the site of genocidal formations that seek to destroy rather than to articulate (even as subordinate). While Walkerdine shows how (white) girls are constructed as having a kind of inauthentic reason, educational institutions have historically excluded students of color according to a logic that has often understood them as not having a human mind at all (Stoskopf 1999). Against this (discursive and material) formation of white supremacy, an ability to locate the subject of domination—the oppressor—in whose systematic interest this violence takes place, and the capacity to invent forms of resistance against it, are matters of survival. Conversely, the unremitting theoretical problematization of such categories may not always be subversive but in some instances may represent the affordance, in theory, of an unacknowledged privilege, as postmodern and poststructural analyses ignore familiar and more direct forms of violence and control in education (Schutz 2004).

Discussion: Reconceptualizing Oppression in Educational Theory

In the face of a widening crisis, a coming together and thorough thinking through of critical tendencies in theory and practice is more and more necessary. While radical activists, theoreticians, and educators may not have found an adequate meeting place in their work, the forces that organize oppression globally are more and more converging into a recognizable center (Gowan 2003) and into an intensified reactionism as well. For teachers, other educators, and activists in the context of this contemporary "emergency time" (Giroux 2003), the kinds of problems and contradictions that emerge from an analysis of oppositional discourses in education are very important, and the need to challenge these different perspectives together toward a collaboration is urgent. This has to be more than a simple juxtaposition. We need an understanding that proceeds from the particular situations of diverse constituencies and conceptions toward a politics these positions cannot contain by themselves—a vision that is willing to risk transgressing the purity of political identification for the sake of a praxis that is richer and stronger.

The Limitations of Standpoint Theory Proper

Part of the reason for the contradictions between the educational paradigms described above, of course, is that they are anchored in different social constituencies. It is not surprising that each appears to be only partial, given that each constructs a viewpoint from a different sociological referent. The primary concern of Walkerdine, for instance, is with the experience of girls, whereas that of McLaren is with working-class people. It can be argued that the different perspectives I have discussed in this chapter are not merely different views, but different deep readings, the products of essentially different political and epistemological standpoints that are themselves created ultimately out of different social experiences. Thus, the problems in each account are not just incidental but rather are the result of their very integrity as analyses with a characteristic shape, mission, and set of problems. If each of these standpoints appears—by itself—inadequate with regard to understanding oppression, this is because those who are controlled and marginalized in schools belong to all of the reference groups assumed by the three theorists considered here, sometimes at the same time. In other words, students in schools are oppressed variously on the basis of race, culture, class, gender, and other factors, and some students are oppressed on the basis of all of these factors at once. Many low-income students, for instance, experience additional struggles due to the way that they are racialized by schools and society; some of this group, in turn, must also contend with marginalization based on their status as immigrants. This complexity is not an overwhelming problem if we only want to describe the different struggles that students face; but if the goal is to develop a coherent political and philosophical standpoint that corresponds and responds to their experiences, then the different standpoints constructed on the basis of a prioritization of singular forms of oppression will necessarily fall short. For example, a political philosophy built essentially on the experience of race and racism will not necessarily provide a foundation for combating the sexism that pervades curricular materials and instructional practices. What is needed is a complex or compound standpoint that can be sensitive to and proceed from the range of struggles that organize schooling.

One potential way out of this dilemma is to recognize that if a standpoint that emerges from a singular form of oppression is inadequate to the different dimensions of students' lives, nevertheless a different standpoint might be constructed that responds specifically to the experience of multiple and intersecting experiences of oppression. A path in this direction is indicated by feminists who have complexified the original formulation of standpoint theory, such as Patricia Hill Collins (2000) and Chela Sandoval (2000). For example, Collins has described the way that Black feminist consciousness cannot be assimilated to that of white women but depends for its essential dimensions on the particular experience of Black women. Black feminist thought foregrounds the "interlocking" nature of oppression, which responds to the actual experience of Black women who have suffered through racial, gender, and economic oppression at once. It also challenges the privileging of sexism as the foundational experience for all women (and the idea that this sexism divides a continuous and unproblematically homogeneous category of women from men) that characterized the early work of many white feminists. Collins' work is essential, in particular, for building an analysis that can respond to the experiences of girls of color in school, since these experiences will be misunderstood by any perspective that focuses on only one dimension of their identities. For example, feminist efforts to support the self-esteem of adolescent girls may fail to the extent that they ignore the different factors that affect white and non-white children differently—and in particular the fact that for girls of color one of the central elements to consider in this regard is racism itself. Likewise, Collins' perspective can allow us to see how dominant educational policy, such as the No Child Left Behind Act, reinscribes dominant forms of knowledge and learning that simultaneously reinforce elite privilege, marginalize students from non-dominant cultures, and valorize a masculinist orientation to truth (that understands it as purely objective, universal, and quantifiable). It could be argued that the complexity of these effects is only visible from a perspective that starts from the interlocking nature of oppression.

However, Collins' account, like other standpoint theories, contains its own problems. Even if it is complexified, it still does not escape the basic problem in early standpoint theory to which it is partly a response: the problem of essentialism. If white feminists were guilty of essentializing the category of women in a way that concealed the deep divisions and differences between women, Collins' paradigm risks doing the same thing in its own way, since the idea of *a* Black feminist consciousness also risks essentializing and homogenizing the experiences of different Black women. This paradigm may depend just as much as other forms of standpoint theory on the construction of a singular subject from whose vantage point the truth of the social becomes clear.[3] With regard to education, one could point to the difficulties in this analysis of comprehending the diversity of experiences of girls of color in schools, as well as the experiences of other variously oppressed students. In addition, standpoint theories generally (including Collins' perspective) can be charged with an idealism that thinks about social truth and social change primarily in terms of the *consciousness* of oppressed groups. Feminist standpoint theorists assume the centrality of epistemological questions to politics; a feminist transformation of the terms in which knowledge can be produced and validated is too often taken as synonymous with fundamental social transformation. But if Nieto, McLaren, and Walkerdine each appear to be bound to different and partly incompatible epistemologies, and if the terrain of knowledge is taken to be the key to social change, then the prospects for bringing these accounts together toward educational transformation are limited. What we need instead is a way of attending to the complexities of oppression and liberatory struggle that nevertheless does not reduce the problem solely to the correct naming of the subject and the articulation of that subject's proper consciousness.

Toward a Compound Standpoint: An Expanded Materialist Analysis

Feminist standpoint theory has wrestled with the problems of multiple and simultaneous oppressions in a way that is very useful for thinking about the problems raised in this chapter. Nevertheless, this tradition runs into a number of obstacles, which I describe above. I believe, however, that these difficulties can be usefully responded to on the basis of an analysis that conceptualizes the different registers of social life as complexly and *materially* linked in history. Starting from the work of Frantz Fanon, this expanded materialist framework understands the economic, social, psychic, and cultural dimensions of oppression as organically interconnected within the problematic of colonialism. Oppression, for Fanon, expresses itself as a "violence which has ruled over the ordering of the colonial world, which has ceaselessly drummed the rhythm for the destruction of native social forms and broken up without reserve the systems of reference of the economy, the customs of dress and external life" (1963, 40). The advantage for educationists of a recognition of this overarching and total reconstruction of social life by oppression, relative to traditional standpoint theory, is that we can begin to understand the links between students' lives and larger historical passages and economies of domination and resistance rather than isolating the particular experience of individuals or groups (and the corresponding forms of knowledge and consciousness) as the privileged ground for analysis and action.

Fanon demonstrated the complexity of colonial society, in which the centrality of class struggle was displaced by the phenomenon of racism. In this context, life chances, wealth, agency, and social value were in the first instance decided by racial assignation. In this society, "you are rich because you are white, you are white because you are rich" (1963, 40). At the same time, however, class relations were present in the context of a capitalist economy and bourgeois society imported from Europe. This meant, on the one hand, that working-class solidarity could not extend in any unproblematic way from Europe to the colonial periphery, since Europeans of all classes benefited from the racist subjugation of Africans. On the other hand, Fanon showed that in the course of struggle, people had to learn that the Black bourgeoisie did not share the same interests as the poor and conversely that there were Europeans who could act as allies in the struggle. In short, to understand the phenomenon of colonialism a complex analytic was necessary that could trace it as a class-cultural process, not only in terms of the experiences of individuals but in terms as well of the reproduction of social structure. In the United States of the early twenty-first century, characterized by an objectively neocolonial organization of educational opportunities, such an analysis is critical to understanding the deep processes that organize the experiences of students.

Later cultural studies scholars, such as Homi Bhabha (1994), have built on the work of Fanon to demonstrate the dynamic hybridity of racial and cultural identifications as discursive processes. What has often been lost, however, is Fanon's commitment to a materialist analysis that understands race and class as more than discursive—as the effects and instruments as well of the economy and political sovereignty. Anticolonial struggle therefore had to involve the concrete appropriation of the wealth and historical potential of the nation for the people—"the realization by the colonized peoples that *it is their due*" (1963, 103)—rather than merely a rearticulation of identities or practices. And in an elaboration of a central insight of Marxism, Fanon showed that the dynamism of social identifications is the effect of their being rooted in the contradictions of concrete social struggle. This can be seen in the way that one-dimensional and Manichean constructions of black and white that dominate the situation of colonialism give way to less essentialist and more contextual understandings of political identity during the movement for liberation. In the context of educational struggles around resources or curriculum, this realization makes possible essential solidarities across class, race, and other social divisions, solidarities in which participants' work is grounded in a commitment to a politics that articulates a collective project and aspiration for transformation.

Fanon's work can be criticized for failing to see the complex ways in which gender organizes forms of oppression, as well as the possibility of liberation, as Anne McClintock (1997) shows. If we are to understand how oppression operates in education and society and the possibilities for praxis against it, we will need to understand how it is gendered and how this gendering constructs its fundamental meanings. Chandra Mohanty (2003) exemplifies such a reinvention of thinking about decolonization. She articulates a feminist perspective that is at the same time an antiracist and anticapitalist one—and that analyzes how all of these axes of oppression are tied to historical projects of colonization. In Mohanty's analysis, as in the expanded materialist one I suggest here, the various moments of social domination cannot be reduced to any simple and singular principle but nevertheless still form a whole that it is important to theorize. The principle that an overarching historical dynamic organizes political, social, and cultural conditions of life allows for a more holistic approach to analysis and a less contradictory one. From this perspective, the sense of "standpoint" is enlarged beyond the epistemological frame that captures it from Lukács to recent feminist theory; it is less the achieved *product* of experience than the constantly reframed *practice* of analysis in the context of material struggle. It does not reside in any single privileged actor but rather attempts persistently to grasp the complexity of the whole from the multiple perspectives that propose themselves in revolutionary transformation.

Overcoming the Contradictions of Educational Theory

On the basis of a compound standpoint that is materially grounded, we can begin to overcome the contradictions between the different educational paradigms discussed in this chapter in a coherent way. First, if we understand culture as an aspect of material life, as Fanon does, then we can address the gaps that characterize the accounts of both Nieto and McLaren. From this perspective, the historical dynamics of conquest and expropriation are political-economic processes at the same time as they crucially operate in a cultural register. This means that treating power relations in culture as abstracted from political economy distorts the meaning of these relations, but also that focusing solely on relations of production in the economy ignores cultural life as an essential register of these relations.

To take the example of language: the dominance of English and the suppression of immigrant students' primary languages links questions of individual identity and affirmation to struggles over class-racial hegemony. The neocolonial structure of class relations in North America means that struggles over language are racialized and that this racialization operates in a way that guarantees the ongoing material subjugation of non-dominant ethnic groups, individually and collectively (Macedo 2000). In other words, the violence done to individual students' spirits, psyches, and identities represents the same process as the preservation of economic and political privilege for elite groups.

This dynamic cannot be grasped holistically from any perspective that analytically splits the cultural from the economic, as do the accounts of both McLaren and Nieto, each in their own way. On the other hand, to recognize the deep interconnection between these dynamics as different registers of an underlying process can fill out the gaps in each of these accounts, making McLaren's account more truly dialectical through a recognition of the priority of the cultural in history and deepening Nieto's account through an understanding of the necessary link between individual experience and collective class struggle.

Not only does the compound standpoint I am proposing here link culture and class; discourse as well, in this approach, can be understood as representing the materiality of historical violence. From this expanded materialist perspective, Walkerdine's emphasis on the very grammar of schooling as creating a set of impossible conflicts for students (especially girls) is not necessarily in contradiction with the accounts of Nieto and McLaren. What is neglected in Walkerdine's analysis is that this grammar, or discursive structure, is not an arbitrary arrangement but rather part of a larger economy of privilege and oppression. Walkerdine analyzes the hidden itineraries of concepts such as "real learning" and "good student" and shows them to be essentially gendered rather than merely expressions of simple bias. But the hidden determinations of "learning," "good," and related constructs such as "success" and "achievement" are connected to systematic class-racial offensives that are designed to preserve monopolies of privilege that link discourse to life chances, social class assignments, and cultural imperialism at the same time that they organize the order of patriarchy in schooling. For example, while the notion of "good student" makes schooling especially fraught and contradictory for girls (as Walkerdine shows), at the level of the system as a whole this notion corresponds to the construction of the "blue ribbon school," which operates as an indispensable sign, for affluent communities, of their cultural superiority.

On the other hand, Walkerdine highlights the fact that patriarchy has been and remains a crucial dimension of material oppression. An expanded materialist analysis, and a compound standpoint proposed on this basis, can link her insights to those of Nieto and McLaren to analyze, for example, the new forms of authoritarianism that have been produced through recent "accountability" initiatives in schooling.

These forms of authoritarianism renew the legitimation for the screening out of low-income students and students of color through tough "back to basics" curriculum, zero-tolerance disciplinary measures, and high stakes testing. At the same time, they are essentially masculinist and patriarchal formations that depend on a sense of the right and good as the property of the strict father, the image of whom is projected into the institutional authority of the school itself. Crucially, then, the gendered nature of these school processes is not separate from their class or racial dimensions—instead, the very masculinism of these formations is the key to their operation as class-racial offensives. Here again, an expansion of the idea of the material to reveal the instrinsic links between registers of experience allows for a more complete and accurate understanding of oppression. The organization of educational discourse acts immediately and materially on the possibilities for students' learning, identities, and life chances in a way that does not merely reflect but rather repeats in its own sphere the economic and cultural violence that characterizes the history of neocolonialism.

A compound standpoint, conceptualized in this way, can disclose essential meanings of a range of phenomena in education that can be only partially grasped from the vantage point of paradigms that assume singular principles as the essence of the social. Second-generation segregation, for example, in which schools are just as divided (or more) in terms of racial composition as they were prior to *Brown v. Board of Education*, is usually thought of as the secondary effect of the vestigial racism that produces residential segregation. In this way, de facto school segregation is supposed to represent a kind of remnant of social prejudice not quite overcome by the civil rights movement. Although this is partly true, what is masked in this account is the more essential *continuity* between second- and first-generation segregation, both of which accomplish, in their own ways, the renewal of an original project of dispossession of the poor and working classes—classes that have always been racialized in U.S. history. An expanded materialist analysis, starting from Fanon, discloses the historicity of segregation as a repeated moment of social violence and reinscription of social privilege. Furthermore, it is no accident that this ongoing process of exclusion is rationalized and legitimized through a set of patriarchal understandings of students as in need of both constant punishment and constant control. The sense of women as fundamentally deviant, which has constituted the history of patriarchy, reappears as the essential character of Black and brown and "low-performing" students in this discourse. Without an understanding of this deep and complex economy, efforts to address segregation will founder, since they will not recognize the fundamental meanings at work and will only displace these temporarily through the mechanism of instrumental remedies, however broad in scope.[4]

Thought of simply as distinct epistemological frameworks (following traditional standpoint theory), the educational paradigms discussed in this chapter cannot successfully be bridged. But if we understand them instead as partial views or reports on a historical process that encompasses them all, it then becomes possible to combine them as complementary evidences of an overarching social violence. While the paradigms considered here each illuminate one aspect of this process, it is only when these aspects are brought together that a full picture of schooling emerges, as well as the beginnings of an adequate outline of social oppression. The logic that comprehends this complex economy must be able to describe, as the approach I have proposed here does, how different forms of oppression are braided together, how they make use of each other opportunistically, and how they collaborate within a neocolonialism that institutes conquest and domination as the inner truth of social life. Only on the basis of such a synthetic analysis can critical and liberatory efforts recognize the proper scope of the tasks they face, and the kinds of strategy that are necessary in response.

Conclusion

The compound standpoint described in this chapter can allow us to make sense of the insights of critical pedagogy, feminist poststructuralism, and multiculturalism together—how to respond at once, in a theory of oppression, to the different social processes that these approaches point to. Rather than simply focusing on each different form of oppression in turn, we can look to the shape of the dominative movement that operates through each of these terms. Against the reification of oppression as a set of distinct and unique forces (operating on correspondingly unique identities), we can begin to recognize the common material and historical basis of these forces in a manner that resists their fetishization as simple objectivities and points to the complex ways that they always inhabit each other. In the field of education we should be especially sensitive to and prepared for such conceptual transformations, to the extent that we are concerned with a form of social activity oriented toward change and development. We should be prepared to think creatively about how different critical accounts can comment on, inform, and combine with each other toward new and more powerful understandings. Education is implicated in cultural, political, economic, and discursive processes at every level—from the global to the "personal" (an opposition that the analysis above already significantly unravels). Against the emerging schema of processes of oppression on a world scale—visible in the global adventures of neoliberalism, as well as in the local reconfiguration of literacy practices—the compound liberatory standpoint described here needs to propose itself over the same difficult terrain, and educators and theorists need urgently to consider what part they can play in its development.

CHAPTER 4

Clearings and Enclosures
Primitive Accumulation and Contemporary Schooling

The beginning of the twenty-first century has seen a remarkable and unsettling convergence of trends in public education. At the same time that public schools are attacked and undermined, and as the movement for privatization has taken hold, the teaching and curriculum within these schools is reorganized within a deeply instrumentalist framework. Likewise, at the same time that the process of learning is rewritten as a simple game of success or failure, the meaning of educational authority is rearticulated in terms of the discourses of the criminal justice system. In this way, the educational mission is reconceptualized as the monitoring and exposure of offenders rather than the intellectual and emotional development of children. All of these tendencies are involved with new forms of tracking, which create different pathways for students within the same regions and districts, and even the same schools. Furthermore, this tracking participates in the complex reorganization of the economy and social life that is brought about by neoliberalism. In this process, racialization and racism, as well as class stratification and class hegemony, are renewed and reinvented according to the contemporary requirements of power and capital. Under the sign of the standardized test, educational opportunities can be tied to the geography of gentrification, apartheid schooling can be naturalized within the logic of "achievement," and a surplus population can be pushed out of the schools and into the wilderness of "globalization."

Understanding this convergence—what these disparate tendencies have in common, and the logic that drives them—is important for those who would hope to struggle for a different kind of education. In the previous chapter, I consider several leading theoretical paradigms for understanding oppression within education; in this chapter, I analyze specific examples of this oppression in terms of prevailing tendencies within contemporary mainstream school reform. I present an analysis that aims to contribute to uncovering a common meaning to contemporary trends in education. Starting from the position that in important respects contemporary economic and political processes echo a much earlier phase in the history of capitalism[1]—namely, the stage of its preparation and emergence—this chapter analyzes current trends in schooling in light of Marx's own analysis of this early period. Marx called this historical stage *primitive accumulation*, and he described it as making possible the birth of capitalism and its development on a grand scale. This stage of capital accumulation was characterized not by an established system of wage labor but rather by a process of plunder and violence, which created the initial store of wealth upon which the first entrepreneurs depended. I return in this chapter to Marx's original narration of this passage. It is my view that his analysis sheds important light not only on present day capitalism as a whole but also on contemporary education in particular, and that it exposes the participation of schooling in the repetition of a very ancient and yet very contemporary form of social violence. This violence, which tears human beings from the very conditions of authentic creativity and development, founds capitalism as a social logic at its inception, and repeats this constitution in the social processes of the present. In the experience of contemporary assaults, human beings relive the trauma that originally reorganized society around the principle of exploitation. Since education expresses those meanings that are central to any society, students and teachers are at the heart of this experience.

Marx's analysis of primitive accumulation also reveals the symbiotic relationship between capitalism and colonialism, a relationship that is alive and well in the present, as new forms of exploitation and new forms of predatory racism reinforce each other within a larger global project of subjugation. This is clearly evident in the arena of schooling and supports my argument throughout this book that contemporary "economic" and "cultural" oppressions ultimately express the same underlying dominative logic. In what follows, as the framework for my discussion I explain what Marx meant by primitive accumulation; I then survey recent educational reforms in the context of capitalism; finally, I describe how Marx's account of the origins of this mode of production helps to illuminate the essential political meaning of these educational reforms.

Primitive Accumulation

The educational trends and tendencies I describe in this chapter are part of a larger political and economic context that is marked by the aggressive encroachments of capital on the public sphere and a disciplinary logic that punishes all who resist its advances. This reorganization of social life, which partly signals the desperate response of elites to an implacable systemic economic crisis and partly marks the ramification of capitalism on a properly global scale, is the project of neoliberalism. Neoliberalism demands the subordination of public needs to the needs of capital and opposes forms of regulation and protection that lower the rate of exploitation of workers and hinder the appropriation of natural resources by private firms. Furthermore, where neoliberalism proper, as a set of economic policies and imperatives, has not been able to procure the required return to global financial elites through the disciplining of economies by international financial institutions, the capitalist state has not been afraid of returning to more directly imperialistic practices both to open up new spaces to investment and to defer its own political crises, as the present conflict in Iraq demonstrates. As David Harvey (2003) shows, these economic and political trends indicate a return to a form of accumulation of capital that proceeds not simply through reproduction on a wider and wider scale of capitalist industrial production but in addition through the violent seizure of public resources and their conversion into the property of transnational firms. The privatization of basic services like water and electricity, the theft of indigenous knowledge (for example by pharmaceutical companies), and the destruction not only of national policies but also of entire states that stand in the way of the penetration of capital are all examples of what Harvey has called "accumulation by dispossession." It is in this context, as Saltman

(2007) has usefully described, that we can make sense of present trends in the educational sphere that aim to open public schools to privatization, tear down the barriers to commercialization and commodification of knowledge, and in general reorient the understanding of learning and assessment, as well as the allocation of resources, in response to the requirements of financial elites.[2]

However, it is important to recognize that current globalizing processes reprise a much older chapter in the history of capitalism. Harvey's insight into the particular aggressiveness of contemporary modes of accumulation is inspired by the account that Marx gives of the initial establishment of capitalism itself. According to Marx, in order for capitalists to concentrate enough wealth and means of production to establish industrial production on a wide scale, and in order for there to be a population of wage-laborers ready (and compelled) to go to work for them, a systematic plunder of communal resources had to take place, both within Europe as well as in its colonial possessions. This process of primitive accumulation,[3] which sets the stage for the further development of capitalism by converting into capital both the natural and social wealth of a people (as means of production), as well as the inherent creativity of human beings (as labor power), is echoed in the present-day pillage of national and communal property worldwide. Contemporary dispossessions seek to respond to current crises of overaccumulation through a return to the brutal methods that marked the dawn of the capitalist era. Revisiting Marx's account of primitive accumulation is therefore useful for the light it throws on contemporary processes; in particular, there are striking resonances between his narrative and current dimensions of the reorganization of schooling.

As Marx describes, capitalism is founded on a grand theft beginning at the end of the fifteenth century, as landed proprietors broke free from the constraints of law and custom of the feudal era and sought to appropriate for themselves the property that had previously belonged to the state or directly to the people.[4] In England, which constitutes the essential case study for this process, arable land that was farmed collectively by peasants was seized by renegade sectors of the nobility and converted to pasture. This is the archetypal case of "enclosure," in which the communal land of the village (the "commons") was sealed off and made the private property of wealthy sheep farmers. In what can be seen as perhaps the original moment of gentrification, the peasant population of the villages was cleared from the land and forced to join the growing ranks of the proletariat in the towns. The feudal obligations between the lord and his retainers were dissolved, and the common folk were "set free" to find their way as best they could in a new system in which all that they owned was their own power to work. Independent peasants were forced to become wage-laborers for capitalist farmers, and individual household industry was destroyed, as producers were forced to work together in factories owned by others and with tools and for purposes that were not their own. At the same time, the wealth of the land, plundered by the new entrepreneurs, became the original capital that allowed for the reorganization of production on a large and coordinated scale.

This process of dispossession established the basic social relationships that constitute capitalism. As Marx explains, its essence is in the fact that it separated "the worker from the ownership of the conditions of his own labor"; in other words, it resulted in "divorcing the producer from the means of production" (1867/1976, 874–75). On the one hand, all that went into the material production and reproduction of society as "means"—the land, tools, and knowledge of the people—was converted from communal resource to private capital. On the other hand, the original producers themselves (the peasants, weavers, and spinners) were torn from the material conditions of work that had belonged to them. They were forced to sell their own capacity to labor to those who now owned the means of production and thus to enter into the ranks of the modern working class, or proletariat. It was this original separation of producers from means of production that allowed capitalism, as the relation of exploitation between capital and labor, to truly take shape. This was hardly a peaceful process, since the people did not willingly leave the lands and dwellings that had been their own and did not eagerly submit to the discipline of the factory. In addition, in its early phases, capitalist manufacturing could not absorb the surplus population of workers that was produced by enclosures and clearings. As a result, terrible laws were promulgated against the displaced who did not find employment. Marx describes how, in accordance with a series of edicts, this surplus population of "vagabonds" was whipped, tortured, and executed in an effort to enforce the new discipline of the factory. We can recognize in these original brutalizations not only a phase in a historical narrative but also the precursors to contemporary forms of criminalization of youth and the poor.

The violence of the dawn of the capitalist era was not confined to England and Europe. A large part of the wealth that was converted into the founding stock of capital was derived from the plunder of the rest of the world by European powers. From this perspective, the viciousness of the slave trade and the pillage of precious minerals and products from Asia and Africa were essential chapters in the establishment of capitalism. Not only did the profits derived from this looting return to European seats of power and commerce to become the source of credit for capitalist enterprises in Europe; in addition, new markets were forcibly created in the colonies for the products of the new manufactures. Marx shows how the capitalist system, at its very birth, is a product of the murder, enslavement, and robbery of non-European peoples, without which it would not have accumulated the store of wealth that served as the foundation for its development and flourishing.[5] Present-day neoimperialist projects that seek to capture globally strategic assets and regions make clear that from its origins up to the present capital has depended upon a racist cultural and economic machinery to procure for itself, on a global scale, the resources that have allowed for its universalization.

School Reform and Capitalist Culture

In this section, I place contemporary developments in school reform in the United States in the context of transformations in capitalist society and culture, particularly as they are affected by globalization and the new forms of discipline associated with it, as well as making preliminary links to the process of enclosure described above. This contextualization will set the stage for the discussion, which follows, of the relationship between processes involved in primitive accumulation and present-day educational reform. In this way, trends in education will be seen both to persistently reproduce the basic attitudes and ideologies required by capitalism, as well as to *reconstitute* it through a repetition, on very different terrain, of the violence through which it first emerged.

Capturing the Curriculum

Recent decades have seen the explosion of test-based "accountability" initiatives, which have not only reframed educational assessment but have also, and perhaps more importantly, led to the wholesale reorganization of curriculum. While currently the No Child Left Behind Act aggressively presses this trend ever more deeply into public-school culture, it is important to recognize that the promotion of norm-referenced standardized tests, as linked to the development of standards, has been gathering steam for some time. Given its initial impetus by the publication in 1983 of the report *A Nation at Risk*, which blamed the failure of the educational system for a decline in the global competitiveness of the United States, this movement was given a boost in the Clinton administration with the signing into law of the Goals 2000: Educate America Act, which supported the development of "standards-based" education reforms. While critics give much attention to the tests themselves, it is important to recognize that the true cost to curriculum and instruction comes also from the extensive test preparation regimes that districts institute in order to avoid humiliation when annual scores are reported. Test preparation reduces teaching and learning to a matter of procedure, squeezing out substantive attention to higher order and analytical thinking, creative subjects and skills, and the possibility of critical or social justice–oriented curriculum (McNeil 2000). Sapping the energy of both students and teachers, it contributes to attrition in the teaching force and to the dropping out of students who are demoralized by the experience both of the testing itself and the shaming associated with it. In fact, evidence suggests that in some schools low-performing students may be deliberately pushed out in order to raise scores (McNeil 2005).

Much of the criticism of standardized tests has come from a traditional progressive-educational perspective, which recognizes in their proliferation a threat to the possibility of experience-based and project-oriented curriculum, as well as to the development of a culture of civic engagement in schools. But the testing movement is also part of a broader assault on schools as locations of cultural negotiation and the production of hybrid knowledges. Not only, for instance, are English language learners often required to take tests in English, a language they do not fully understand; in addition, the narrowness of the senses of achievement and learning that are inherent in the logic of norm-referenced testing reinforces a subtractive understanding of education and student culture more generally (Alamillo et al. 2005). The proliferation of scripted curricula, which reduce instruction to a "teacher-proof" and instrumental process, accompanies the narrowing of the meaning of assessment, and likewise circumscribes the ability of educators to extend the meanings of learning to include complex multicultural and critical terrains. Gutiérrez et al. (2002) are right to note the links between these developments in education and broader attacks on people of color and Latino/a immigrants in particular; they describe this ensemble of policies and cultural trends as a "backlash pedagogy" based on a politics of white middle-class resentment that seeks to preserve its privileges in the context of a diversifying society and an uncertain economy.

The transformation of curriculum and instruction according to a reductionistic obsession with test scores marks the penetration, at many levels, of the logic of capital into the public space of education. This is a contemporary expression of what Marx (and earlier observers) described, with regard to the original appropriations by capitalists of public land and resources, as a process of *enclosure*. As many have noted, the restructuring of curriculum has as one important effect the corporatization of a much larger proportion of educational life. States contract with private corporations to acquire not only standardized tests but also packages of scripted curricula, as well as computerized reading comprehension assessments. The test production and scoring industry alone generates hundreds of millions of dollars annually in sales, attracting eager start-ups to compete with the dominant players, with the race for profits leading to frequent errors in test design and scoring (Gluckman 2002). The systems of external evaluation, remediation, and supplementary educational opportunities that accompany test-based accountability are further opportunities for the penetration of capital into the educational "market." And of course, the testing movement can also be seen as a long-term strategy to demonstrate the failure of public schooling more generally and to open the door to charters and vouchers (Valenzuela 2005). At the same time, these initiatives inculcate an ideology of submission to the rhetoric of business in the articulation of student and school assessment in terms of benchmarks, annual goals, and the bottom-line logic of score-reporting, which are understood on analogy to the quarterly earnings reports of corporations.

But the capturing and enclosing of collective life that is at work in this reframing of education goes even deeper than these analyses suggest. Not only does the reduction of literacy and learning to the absorption by students of a list of testable items open the door to a spectacular set of profit-making opportunities for private corporations; in addition, in this process knowledge itself is recast as a form of alienated property. Students are given the mandated materials and directed toward the demanded results according to a logic that is alien to the natural course of their own learning. Cognition, imagination, and the very souls of students are expelled from the terrain of schooling to the extent that they do not conform to the shapes and modes that official instruction anticipates. This is especially clear in the case of students whose cultures, languages, histories, and identities differ from or even contradict the logic of the official curriculum and who are thus implicitly written off as illegitimate, inessential, or even criminal in their difference from what is presented as the universe of the right and the normal.

Hyper-Disciplinarity

Deeply connected to this reorganization of curriculum is the intensification of disciplinary processes in schools, as a result of which education is increasingly experienced as a form of punishment. This development extends into schooling a deep-seated trend in society generally to criminalize youth, and in particular youth of color, through the passage of antiloitering ordinances, gang injunctions, and mandatory sentencing minimums, as well as the practice of trying child offenders as adults. In schools, students face intensive surveillance, random drug searches, zero-tolerance policies that expel them for nonviolent infractions, and in general beefed-up security apparatuses. Draconian disciplinary policies that result in increased expulsions disproportionately target African-American and Latino/a youth as school discipline increasingly takes on the form of law enforcement and in some cases is actually undertaken by the police themselves (Dohrn 2001). This trend toward the criminalization of students coincides at the same time with a divestment from their authentic academic and emotional development and a systematic abandonment of the responsibilities of educators as mentors for young people (Devine 1996). This can be viewed at once as both the beginning of a process of social marginalization for the poor and those who are not white (leading in many cases to the prison system itself) and at the same time, for those who are able to accommodate themselves to these regimes, as a kind of militarization of subjectivity (Lipman 2004)—not to mention an actual preparation for the military itself, which as a result of No Child Left Behind has increased access to the crucial recruiting ground of high schools.

This carceral trend in schools influences the meanings of curriculum and teaching. Not only do the rituals of testing become a form of moral examination (determining, in popular parlance, which are the "good" and "bad" schools); in addition, retention policies and high school exit exams act explicitly as immediate arbiters of the virtue of students and of their right to a share in the goods that credentialing makes available in a cutthroat economy. Anagnostopoulos (2006) shows how teachers, through a complex moral discourse that obscures the collusion of retention schemes in larger processes of social reproduction, seek to rationalize the promotion or demotion of students and to distinguish between students who "deserve" their academic fates and those who have been perhaps set back temporarily. Similarly, the liminal space of detention, a kind of midway point between an academic and a juridical response to minor offenses in the classroom, is increasingly fortified into an actual form of imprisonment within the walls of the school (Lyons and Drew 2006). Many have pointed to the remarkable coincidence between these contemporary forms of school discipline and assessment and the organization of modern society in terms of the perpetual surveillance of the panopticon that Foucault described (De Lissovoy and McLaren 2003; Devine 1996; Vinson and Ross 2003). However, surveillance functions differently in different spaces. In suburban schools, as Lyons and Drew show, surveillance operates not simply to construct students as docile subjects but also to demonstrate the racial difference between suburban and urban space. As they describe, lockdowns, security cameras, and drug searches operate in the middle-class white environment to exorcise any traces of difference—to show, in the very triviality of the infractions detected (e.g., students skipping class), that the suburban school is not the dangerous and chaotic place that the spaces of urban schools are taken to be. On the other hand, to the extent that intensive surveillance is part of a disciplinary system that is characterized by high rates of retention, suspension, and "dropping out" for students of color in urban systems, it is important to emphasize the very material effects that it has on their occupational and life chances.

One important effect of these disciplinary trends in schooling is in fact to aggressively discourage students from being successful. In this regard, the new accountability also has to be seen as a strategy not simply for organizing the entrance of students into a new and highly stratified global economy but also as a means to exclude them from participation in the first place. In this pattern, we can recognize the imperative to maintain, now on a global scale, what Marx called an "industrial reserve army"—the surplus population that is always present as a reserve to be absorbed into the workforce as capitalism expands, and that acts at the same time as a pressure to keep wages down for those who are employed. In the context of neoliberalism, in which firms compete to draw the greatest possible surplus from more and more intensively exploited populations, this is a real imperative, especially in "core" regions, which have historically been protected from the worst ravages of the global race to the bottom. This principle aggressively expels individuals from participation in production and civic life, encircling the space they inhabited as the property of the privileged. In this sense, the literal gentrification of urban space is reproduced figuratively in the diminishing access of low-income students to an authentic educational experience and to the currency of educational credentials. At the same time, the targeting of students of color for discipline and expulsion is implicated in the formation of a new global system of apartheid (Hardt and Negri 2004), which flies under the radar, since it appears as the side effect of apparently "objective" social and economic processes rather than being the result of any *de jure* discrimination.

Resegregation, Retracking, and Gentrification

The resegregation of students, which began almost immediately after the first achievements of integration following *Brown v. Board of Education*, even more directly contributes to the construction of an apartheid system in the United States. Scholars have noted the shortcomings and limited effects of *Brown* and have admonished activists not to rest in a nostalgia for a partly imaginary moment of social progress (Ladson-Billings 2004). Nevertheless, schools in the South were integrated to a significant extent in the wake of this decision, and it became in addition a crucial symbolic gain in the struggle against racism. However, in recent decades these gains have been eroded. In the present, in many places schools are as segregated as they were thirty years ago. At the same time, the majority white population does not experience the same outrage at this situation that it did a generation ago and has been content to leave students of color to the mercy of schools that are substandard (Kozol 2005). Instead, activists and attorneys have tried to make use of state constitutions and courts to guarantee a minimum level of adequacy for those in the schools that are the worst off rather than arguing the case for equality of conditions, not to mention substantive integration. But this strategy abandons the broader struggle against systemic racism and leaves reform efforts at the mercy of a patchwork of different decisions for different jurisdictions (Street 2005).

Resegregation, as well as the retracking of students within schools, is connected to deep transformations in the economy. Pauline Lipman (2004) argues that racial isolation in the public schools is connected to a global economy that must aggressively stratify workers into managerial and service positions and in which processes of urban gentrification displace the poor in order to attract wealthier residents to neighborhoods and schools. She shows that new college preparatory programs in Chicago, highly touted as part of educational reform efforts in the city, are situated in areas currently undergoing gentrification and are designed to appeal to these new residents. The tracking that these programs produce internally is ultimately driven by the demands of business groups eager to position the city competitively, in a globalizing economy, as an attractive place for affluent people to live and work. Contemporary developments in education thus reveal the way in which processes of racial and class stratification coconstruct each other. Part of the inability of the discourse of racial equality to mobilize outrage against the disparities in opportunities between white students and students of color is that that racial segregation has been understood as the arbitrary effect of economic imperatives that are thought of as "objective" and inevitable. Of course, the neoliberal logic of "globalization," which is often imagined as a working through of colorless contradictions in the economy, is a deeply racialized and racist project. Haymes (1995) describes how urban gentrification fundamentally aims to displace Black people from the city and to disrupt Black urban popular culture. Indeed, race and racism are important dimensions of global processes of exploitation and marginalization.

One consequence of globalization is that urban social polarization is reorganized along transnational lines. At the same time that forms of legal discrimination and segregation within nations such as South Africa and the United States have been dismantled, new processes of stratification are introduced that operate on new global actors across national boundaries. According to Sassen (1998), "global cities" produce a characteristic polarization between managers of the transnational economy and the service workers who fulfill the needs of the urban elites. Both of these populations are increasingly "cosmopolitan"; for instance, the armies of custodial and domestic workers in Los Angeles and New York are increasingly made up of immigrants. At the same time, those who were subject to systematic discrimination and exploitation as part of specific *national* dynamics, such as African Americans, may find themselves *re*marginalized as an effect of globalization. And in this new context, the old tools (e.g., litigation) may yield only minimal results. Anyon (2005) has argued that the deep problems of public education can only begin to be remedied as part of a broader reform of public policy in the United States that recognizes the links between education, housing, transportation, and the system of taxation. However, this sensitivity to context must extend even more broadly to include the horizon of the global and the necessity of struggle and transformation at this level.

"Liberating" Students: Choice and Privatization

Perhaps the clearest expression of the capitalist logic of enclosure in contemporary schooling is the trend toward privatization. The institution and practice of education, from preschool to higher education, represents a vast terrain of human production and creativity. There is perhaps no more tantalizing prize for capital in the present than this terrain, if only the old-fashioned attachment of schooling to ideas of the public good and the responsibility of the state, not to mention the legacies of progressivism, can be broken. In education, privatization includes the proliferation of for-profit charters as well as voucher schemes. It also includes the tendency toward commercialism (Molnar 2005), in which corporate brands are given privileged access to the space of the cafeteria, the classroom, or the textbook. As Saltman (2007) describes, while on the whole the privatization movement appears to be somewhat on the defensive, nevertheless the corporatization of schooling is increasingly taking place through the back door, as crisis, disaster, and school "failure" become the occasion for an opportunistic seizure of the public space of education.[6] As he shows, the collapse of the school system in New Orleans in the wake of Hurricane Katrina has become an opportunity for a tremendous experiment in privatization, as the regular public schools have become almost entirely replaced by charters and as federal money is pumped into vouchers as part of "emergency" aid. While the privatization movement presents itself publicly as the route to greater empowerment for the most disenfranchised students and communities, privately its advocates are frank about the tremendous financial returns potentially to be realized (Kozol 2007). Likewise, the accountability movement can itself be analyzed as an enormous Trojan horse designed to pry open the doors of the schoolhouse to private capital, as the humiliation of low-performing schools sets the stage for radical choice-oriented options. This is certainly the case with the No Child Left Behind Act, which contemplates a progressive peeling away of the "worst" schools by charter initiatives, including for-profit ones, if these schools fail to raise their test scores sufficiently.

Not content with the pervasive pushing out of students from the schools through the medium of retention policies, over-testing, and high school exit exams, the forces of choice and privatization seek a wholesale clearing of the population from regular public school systems. Private voucher schemes propose an emptying of the traditional schoolhouse and the dispersal of free and atomized individual students into the "market." Even public choice systems, it is argued, must be "almost beyond the reach of public authority" (Chubb and Moe 1990, 218). Unencumbered by the distortions of social guarantees and expectations, this market and these new educational consumers can then make the perfect union, as students and parents exercise their rational choice in selecting the best school available. Never mind the fact that choice advocates themselves understand that schools will in fact be able to accept or reject whichever students they wish (Kozol 2007) and that any promissory note carried by these new educational consumers (in private choice schemes) may be far from covering tuition, let alone the additional expenses of transportation and supplies. What is important for the choice and privatization advocates is that what could be called the "means of production" of education—the instructional practices, school sites, and curricula—be set free from their obsolete attachment to the state and the people, and that the identities of students as well as the content of democracy be reconfigured away from a civic and political determination and toward a simple economic one (Apple 2001).

It is important to recognize that the "choice" movement is not just about the seizing of public space for private gain. It is also a matter of the basic reorganization of social relationships. Schooling occupies a key place in the process of social and cultural reproduction. Reconfiguring the relationships between teachers and students, and between school personnel and communities, in business terms (teachers as managers, parents and students as customers) prepares the ground for a much broader reimagination of the meaning of social life across many spheres of the public. This is not only because schools are spaces in which students are ideologically interpellated and constructed as subjects in preparation for later insertion in a capitalist economy but also because education is a crucial screen upon which society projects its own deepest self-understanding. Education is the moment in which, as Dewey (1944) explained, society selects from a range of possible senses for itself and chooses a historical trajectory. Expelling democratic and dialogical relationships from the space of school is the essential first step in dismantling the possibility of democracy altogether. For-profit charter systems replace one set of founding principles with an entirely different one. In their very constitution as for-profit entities, these schools express a commitment to the values of individualism, competition, and managerialism, which then influence school organization itself. In the Edison Schools, for example, discipline is based on a "student management plan" consisting of codes for different building areas as well as an "incentive plan" designed to reward positive behavior.[7] Of course, here, as in the workplace, we see the flip side of the values of the entrepreneur in the characteristics of a disciplined workforce: subordination, compliance, and regimentation.

Clearings and Enclosures in Contemporary Schooling

The trends in school reform discussed above should be understood in the context of globalization, and more specifically in relation to neoliberalism and its assault on the public. However, if these educational trends are in fact expressions of this economic regime and its implacable impulses toward expropriation, then it is useful to consider what light the process of primitive accumulation itself, as the founding dispossession of capitalism, throws on them. In particular, while a good deal of attention has been paid to the free-market fundamentalism that underlies the educational accountability and privatization movements as well as global processes of economic structural adjustment, recent school reforms have not been explicitly analyzed in terms of the fundamental secret of primitive accumulation as described by Marx: the separation of the producers from the means of production. Below, I describe and analyze what I believe are the most important connections between Marx's analysis and contemporary schooling. Current educational processes reprise the founding violence of the first capitalists, in which they *enclosed* the common lands of the people, and *cleared* these lands of the populations that had traditionally worked them. Education is today under a threat that is more than analogous to this original plunder; in fact, the aggressive encroachments of capital on the public space of schooling *repeat*, in another time and on another terrain, the acts of theft, annexation, and criminalization that originally gave birth to this mode of production.

Enclosing Education: Corporatization and Alienation

Education is a unique sphere of social activity, since it is concerned with basic social meanings, the construction of citizens, and knowledge production. In addition, in the context of capitalism, it is partly the locus of formation of the essential commodity from which all surplus ultimately derives: labor power. It is important to remember that modern public educational systems are already an expression of capitalist society and its laws. The complex organizations of modern school systems, and the intricate pathways and gateways they propose to students, are already reflections of the division of labor and class stratification that characterize capitalist society (Bowles and Gintis 1976). Nevertheless, the persistence of the provision of mass public education by the state should also be recognized as a prize won and held by working people against their mere reduction to human capital and as an expression of the genuine impulse to democracy that bourgeois society unleashes (and that it can then not fully control). In this sense, public schooling on a mass scale represents a domain that has been held partly separate from the logic of capital, and its fragmentation and privatization does represent an alienation of communal property, or enclosure of the commons, which speeds the subjection of the world to the market and dims hopes for a more authentic democracy.

The plunder of public and communal property in the period of primitive accumulation is remarkably similar to processes of gentrification, neoliberal annexation of state services (including education), and even the opening up of national forests and parklands to industry in our own day. In the period of the first enclosures, common folk were expelled from their homes and livelihoods as land belonging to the state or to the people was absconded with; even Catholic church property was seized and "given away to rapacious royal favorites, or sold at a nominal price to speculating farmers and townsmen, who drove out the old-established hereditary sub-tenants in great numbers, and threw their holdings together" (Marx 1867/1976, 882). In a similar free-for-all in our own time, we have seen the public airwaves, mining and logging rights on public lands, and entire public school districts surrendered to private corporations—not to mention, more recently, the transfer of many military functions, as well as emergency and security services formerly handled by states, to private entities (Klein 2007). Corporations have seized these spaces and resources in the name of efficiency, development, and even libertarianism, while hardly managing at the same time to conceal their outrage that any margin of the store of social value or creativity should resist being converted to the pure purpose of profit-making.

This contemporary pillage, a globalization-era remaking of the social field in the interests and image of capital, echoes the original campaign that Marx describes and should remind us that the process of privatization is anything but peaceful: "The spoliation of the Church's property, the fraudulent alienation of the state domains, the theft of the common lands, the usurpation of feudal and clan property and its transformation into modern private property under circumstances of ruthless terrorism, all these things were just so many idyllic methods of primitive accumulation. They conquered the field for capitalist agriculture, incorporated the soil into capital, and created for the urban industries the necessary supplies of free and rightless proletarians" (1867/1976, 895). What distinguishes our own moment from this original one is that with the capitalist mode of production already having been largely universalized, in the present capital seeks a second-order penetration into social life, seizing even on the deep social resources and capacities of communication and knowledge. This contemporary plunder is represented not only in the commodification of information in the high-tech industries, or of expression and communication in the service industries, but also in the seizure of educational spaces of knowledge production and human development. Like the original dispossession described by Marx, this contemporary one is connected to the displacement of communities. Charter networks, magnet schools, and choice plans help to reorganize the geography of cities, collaborating with gentrification in neighborhoods to remake desirable urban spaces free of the uncomfortable presence of the poor or the racially "different." The parallel should be clear to the original "clearing of estates," in which village inhabitants "were systematically hunted and rooted out" by proprietors intent on raising the profitability of their demesnes, and in which "villages were destroyed and burnt, all their fields turned into pasturage" (Marx 1867/1976, 891). In the name of capital accumulation, which bourgeois reason sanctifies as the first and last moral law in the kingdom of the "free market," actual human beings—whether the peasants of another era, bound to the land over centuries, or the families, children, and students of the present—must be prepared to discover a new "freedom," as development and its engineers liberate them from the histories of place and the privileges of community and cast them into the wastes of a future with no guarantees.

These appropriations and clearings, in schools and curriculum, are blatant injuries to our instinctive sense that education has a larger democratizing purpose. To make sense of these trends, however, it is important to remember what Marx isolated as the secret of primitive accumulation: the separation of the producers from the means of production. In the context he was concerned with, this meant above all the separation of the rural population from the land and from its customary rights to the use of the land. This separation created an original fund of capital for the first capitalists, as well as a population with no other resources beyond its own labor power: the first proletarians. In the context of schooling, the separation of the producer from the means of production means the separation of teachers and students from ownership of the means and conditions of teaching, learning, and knowledge production. It is not only in the breakup of public systems that this process can be observed; the reduction and reification of curriculum, the trend toward partnerships with business, and in general the corporatization of the controlling metaphors of education are all proliferating in public schools. However, in for-profit charters and voucher schemes this essential separation achieves a certain perfection: at the same time that the conditions for learning become the property, not only in form but in fact, of capital, the constituents of the system are remade as a rootless and rightless population at the mercy of the market. While the parallel with the period of primitive accumulation is most obvious in the case of teachers, who become truly proletarianized as the wage-laborers of the new educational plantations, it is perhaps more interesting in the case of students. Students are remade not as workers, but as the perfect consumers, compelled to "choose" a private firm (or privately operated charter) to develop their capacities as pure human capital. Cleared in this way from the domains of the public, ethical, and political, students are allowed to recover their true identities under capital as reified and atomized quanta of strategic and self-interested rationality, reproducing in miniature the core meaning of the institutions in which they are newly educated.

We can also see this separation of producers from means of production in the reorganization of the very meanings of pedagogy, curriculum, and learning. Progressive education from the reconstructionists to multiculturalism has insisted on the idea of students as active participants in their own learning and as co-producers of the knowledge that is made and learned in school. Although not dominant, this sense of education as collaborative and constructivist has deeply influenced schooling. Nevertheless, the contemporary "structural adjustment" of education in the United States seeks to destroy the rights of teachers and students to participate in the construction of educational meanings, and of the meanings of education. At this level, the separation of the producers from the means of production takes on a powerful new significance. Here we come upon the secret of the reductionistic trends in literacy and assessment that confound progressives: parallel to that original separation described by Marx, these initiatives aim to divorce students from authentic involvement in their own learning and then to reinsert them into a terrain in which *their own learning takes place essentially for others, on the terms set by others, and according to a logic in which the initial stock of knowledge, and even that which learning itself produces, always belongs to others.* It is important to recognize that the rituals of high-stakes testing, for example, are not merely exercises in competition and conformity, but that in addition they participate in a crucial dimension of social reproduction: the reproduction of meanings. Forcing students to internalize the compulsion to incessantly produce the right answer is not only a matter of the proper disciplining of social subjects; it is also the incessant reproduction of these answers themselves, the truth and meaning of which belong to the owners of the social. Schools are the factories of this production.

Clearings, Criminalizations, and Neoimperialism

Concurrent with the original enclosures and dispossessions that Marx describes was a process of criminalization of the population that had been driven by these enclosures from their previous dwellings and occupations. This surplus population, set free from the rights and obligations of the feudal period, became the new proletarians. However, the production of this new class was at the same time the production, on a grand scale, of the vagabond:

> The proletariat created by the breaking-up of the bands of feudal retainers and by the forcible expropriation of the people from the soil, this free and rightless proletariat could not possibly be absorbed by the nascent manufactures as fast as it was thrown upon the world . . . Hence at the end of the fifteenth and during the whole of the sixteenth centuries, a bloody legislation against vagabondage was enforced throughout Western Europe. The fathers of the present working class were chastised for their enforced transformation into vagabonds and

paupers. Legislation treated them as "voluntary" criminals, and assumed that it was entirely within their powers to go on working under the old conditions which in fact no longer existed. (Marx 1867/1976, 896)

Penalties for the crime of vagabondage included whippings, brandings, forced slavery, and execution. Anyone not employed and found begging, who was otherwise able-bodied, could be prosecuted under these laws. In advanced capitalism, the universalization and naturalization of the capitalist mode of production mean that workers have no economic choice, and can conceive of no other choice, beyond that of selling their labor power as wage-laborers to survive. At the dawn of the capitalist, era, however, this brutal apparatus of law and torture served to terrorize the new workers into accepting the discipline of the new regime.[8]

In the same way that there are important echoes of the original enclosures described by Marx in contemporary privatization, there are striking parallels between this original production and criminalization of the proletariat and contemporary processes of criminalization in schools and society. Critical sociology of education has historically tended to view the relationship of schooling to working-class students in terms of a process of socialization toward roles within a stratified society and economy (e.g., Bowles and Gintis 1976; Willis 1977). While the functionalist impulses of this tradition have been critiqued (e.g., Giroux 2001), it also appears that the emphasis on social incorporation in these early accounts overlooked the importance of exclusion, marginalization, and violence as modes of social regulation and reproduction. The current intrusion of the discourse as well as the physical apparatus of law enforcement as a core meaning in the space of school links schooling to the prison system literally and figuratively. In this context, the disproportionate referral of students of color for disciplinary action, as well as for special education services, acts to construct and label such students as "abnormal" and to further marginalize them from the center of school culture (Reid and Knight 2006). In addition, at the structural level, education policy as high-stakes assessment, aggressive retention, and back-to-basics curriculum serves to maintain existing inequalities through a process of calculated discouragement and demoralization of students (McNeil 2005). Not only do these policies and practices act as methods of preserving the privileges of white and middle-class communities; furthermore, as a process of production of subjectivity, and as a means of disciplining the working class and the surplus population that is capitalism's "reserve labor army," contemporary educational disciplinary regimes mirror the violence directed at the original proletarians and reveal the often-overlooked fact that violence and exclusion have remained crucial tactics in the repertoire of capitalist hegemony since the period of primitive accumulation up to the present.

Just as we can see a return to the process of dispossession in contemporary capitalism that recalls the period of primitive accumulation, in the present bellicose attitude and practices of the United States we can see a return to the imperialist strategies of the European powers. In the context of systemic economic and political crisis, this new imperialism creates the conditions, under the military leadership of the United States, for the unimpeded accumulation of capital on a global scale. This resurgent belligerency deeply impacts education, as schools are eagerly exploited by the military as recruiting grounds for unpopular wars (Berlowitz and Long 2003). Naturally, the primary targets in this process are the low-income students and students of color whose opportunities are already limited by educational inequalities. Researchers have described the deep militarization of the culture of schools serving African American and Latino/a students, which are characterized by restriction, repression, and authoritarianism, as well as the proliferation of JROTC programs (Brown 2003). More generally, the repressiveness, militarization, and racism of contemporary U.S. society can be understood as part of a global political turn in which capital increasingly turns to coercion rather than consent to maintain its hegemony (De Lissovoy and McLaren 2005). At the same time that the predatory logic of empire spreads outward across the territories of old possessions and new spheres of influence, so too does it fold inward to reframe the meanings of citizenship, democracy, and patriotism and to institute a kind of new and creeping apartheid. In this audacious and paradoxical reorganization, students become both the objects of neoimperialism (as the prey of recruiters and as the victims of a racist and classist arrangement of educational opportunities) and its subjects (as the shock troops of the empire). In the same way that the explosive violence of capital in its infancy terrorized both the English countryside and its global territories, neoimperialism in the present demonizes and destroys both "at home" and abroad.

There might appear to be a contradiction between a more aggressive projection of state power domestically and the antipathy of neoliberalism to state intervention. On the one hand, the state is extending its infrastructure of surveillance, law enforcement, and "corrections," all of which impact the culture and practices of schools. On the other hand, the "colonizations of the lifeworld" (Habermas 1987) associated with the paternalism of the welfare state, which have been described as producing a form of internal colonialism with regard to their management of Black and brown populations (Blauner 1972), appear increasingly to be unceremoniously abandoned by a process of structural adjustment dead set against all forms of social provision. Here too, however, Marx's account of primitive accumulation is useful. As in the case of the criminalization of displaced peasants and the domination of colonized populations in the period of capitalism's birth, the carceralization and militarization of society in the present are not simply the effect of an economic imperative but also express a mode of sovereignty tied to capital as an encompassing logic of the social. In being marginalized in terms of work and public life, people are still included within a total logic of power that depends as much on, and invests as much in, the subjection of bodies before its authority as it does on their incorporation within its economic metabolism. The prison-industrial complex, for example, systematically casts out human beings from participation in social life at the same time that it represents a stupendous investment (in both monetary and moral terms) in the infrastructure and organization of their incarceration. Domestically and globally, capital seizes and then abandons, investing in and then devaluing whole populations, sectors, and societies according to a terrible cycle that should ultimately link struggles against economic exploitation with struggles against social exclusion (Harvey 2003). In this context, the regime of neoliberalism involves both the expulsion of bodies and their simultaneous reinscription within the reorganized order of the social as the perfect subjects, in their spectacular abjectness, of empire itself.

Conclusion

Contemporary challenges in education for progressives (the rise of the voucher movement, the ascendancy of high-stakes testing, the militarization of schools, etc.) are too often thought of as separate. To the extent they are linked, it is usually within a nebulous concept of a resurgence of the forces of the Right. I have argued that returning to Marx's account of the moment of primitive accumulation sheds light on contemporary schooling. In particular, this analysis reveals the correspondence of current trends in education to a number of processes that constitute capital as mode of production and as an organization of social relationships, and that return to reconstitute capital even in the present, in the face of its own inherent contradictions. Above all, the notion of primitive accumulation illuminates schooling as a crucial domain in which can be seen the process, repeated from its origins up to the present, of the *separation of human beings from the conditions and resources of their own creativity and development*—in Marx's terms, the separation of the producers from the means of production. An understanding of this original catastrophe, and the way it is repeated in the present within the fields of economy, politics, and culture, can reveal the organic links between seemingly isolated moments of conservative reform in education. From this perspective, the auctioning off of school districts to for-profit "educational management organizations" on the one hand and the instrumentalization of teaching and learning on the other can be seen as two sides of the same process: the separation of learners from the conditions of learning and the forced rearticulation of both within the logic of capital. In addition, as I have discussed, this return to Marx throws light on the systematic violations that characterize schooling in the United States. The draconian disciplinary policies and the systematic pressure that marginalizes and expels poor students and students of color recall the criminalization and expropriation of the population at the beginning of the capitalist era.

While explicating these educational phenomena in terms of Marx's analysis of primitive accumulation means returning to his narration of the development of capitalism in Europe, we do not need to look any further than the vast theft of land and resources from Native American peoples by the colonists in the "New World" to see the process of primitive accumulation in action. This plunder, together with the enslavement of Africans and their forced labor in the plantations of the colonies, provided the original fund of wealth—as land, resources, and human labor—upon which the spectacular growth of capitalism was achieved during the later history of the United States. Fully understanding contemporary enclosures in U.S. education would involve tracing the continuities and legacies of primitive accumulation in the present day specifically back to these founding crimes of the nation. Furthermore, a consideration of this history also suggests an expansion of the notion of dispossession itself. The expropriation of the lands and lives of the indigenous peoples of the Americas was premised on a religious conquest that defied that which was held sacred by indigenous peoples, and forced them to accept an alien religion that acted as the cultural and ideological arm of the invaders (LaDuke 2005). For this reason, any complete analysis of dispossession must consider the spiritual dimension, and the deep identities and aspirations of cultures, as terrains of struggle and resources for resistance. The limited forms of self-government that have been allowed to Native Americans do not restore the tribes' national and spiritual sovereignty, in the name of which they continue to struggle (Deloria 1984).[9] Extrapolating from this perspective, contemporary assaults within society and education should also be viewed as an effort to dominate the spirit and to control the deepest springs of human meaning within students from which the desire to refuse and to resist might emerge.[10]

In this chapter, I have applied Marx's original insights on the foundation of capitalism to contemporary schooling in order to illuminate the meaning of recent trends in education. This discussion suggests that meaningful change in education will not take place through technical fixes, however broad in scope. While progressives have long recognized the relationship of educational inequalities to processes of social reproduction and to histories of marginalization, my discussion here suggests that the operation of power in schooling is rooted as well in a repetition of the founding violence of capitalism. From this perspective, schools do not simply aid in maintaining the effects of capitalism with regard to class stratification and racial exclusion; they also reconstitute, in a continuous process, the centrality of the deepest meanings of this mode of production—in repeating and naturalizing the violation of human meaning and identity as an essential aspect of the experience of being educated. It is an understatement then to say that we have to deal here with cultural or ideological offensives in schooling in addition to economic disparities. What we struggle over in education are even greater stakes: the deepest sense and logic of an entire form of life. To challenge capital and power at this level means to challenge their very intelligibility. It therefore means, at the same time, to produce a different world and a different meaning for the world. A reinvented conception of education ought to play a central role in this process. In the chapters that follow, I describe the outlines of this struggle.

CHAPTER 5

Difference, Power, and Pedagogy

One of the most important emphases in critical education in the last several decades has been the idea of difference, and cultural difference in particular. While mainstream educational reform in the first part of the twentieth century was a dynamic field, rarely did it confront this issue directly, apart from an attention to the importance of individualizing instruction on the basis of student interests or direction of development. Much of the modernizing rhetoric of progressive education was focused on the construction of a single rationalized civic and social space to which students would be connected on the basis of individual talents, proclivities, and preparation and in which the question of cultural difference was sidelined. In this context, the emergence of the movements of multiculturalism and feminism in education in the wake of the civil rights movement were revolutionary events. While growing in part out of the progressive tradition's belief in the link between education and democracy, these new approaches suggested that the very failure of this tradition to achieve its goals on a broad scale may have been due to its neglect of the issue of exclusion on the basis of difference. These new approaches inherently raised the problem of *discourse*, including the unacknowledged discourses that organized the meanings even of progressive approaches to teaching and learning.

These developments in education were part of a broader shift on the left toward a recognition of the political importance of culture, language, and representation, as well as toward a focus on the liberation movements associated with marginalized identities. The struggles of African and Mexican Americans, women, and gays and lesbians were not simply about widening the social possibilities for oppressed communities but also about recognizing the importance of the political principle of difference itself

(Young 1990). As these movements developed, this insight did too, so that eventually theorists began to foreground the importance not of settled identities but rather of the process of differentiation, and the accompanying phenomena of border-crossing and hybridity (as I discuss further below). In philosophy, poststructuralism sought to expose the violence done to self and language in the enforcement of consolidated and monologic discourses and identities.[1] Much of the criticism that arose from these perspectives, particularly that emerging from the Foucauldian tradition (see Foucault 1977, 1980), was directed not at reactionary or conservative cultural forces but rather at progressive and humanistic ones. It was argued that the latter systematically concealed their own discursive violence against alternative forms of knowledge, understanding, and expression.

These emphases have been crucial in widening the scope of politics and education. They have made possible the assertion of voices marginalized on the basis of race, culture, gender, and sexuality. They have exposed the monoculturalism of the mainstream and have demonstrated the necessity of sensitivity to the exclusions that constitute discourse generally. Nevertheless, the present era has seen the persistence of educational and social inequalities in the United States and an increasing polarization of wealth, standards of living, and life chances on a global scale. Capitalist globalization has discovered new domains of life to penetrate, and new modes of accumulation, while also subjecting more and more people globally to the same economic imperatives and conditions. In this context, it is important that an attention to difference not conceal commonalities in the social conditions of disparate populations. Globalization also shows the continuing importance of economic processes in organizing people's lives and selves and suggests the risks of an exclusive focus on culture (narrowly conceived). At the same time, voices from the global South have interrogated the politics of multiculturalism; postcolonial theorists have rearticulated the notion of difference in terms of the dynamics of imperialism and Eurocentrism (Mohanty 2003), as against an idea of difference based only on discrimination in the global North. In this new conjuncture, a focus on difference is insufficient by itself; critical theory must also investigate the new forms of commonality that construct social life globally, and the possibility of new bases of struggle. The task is not to revert to an era prior to that of multiculturalism or cultural politics but rather to be informed by these crucial moments while building a new global politics and solidarity.

In this chapter, I consider several of the paradigms of cultural difference that are most influential in education and describe the movement in them from an emphasis on affirming the value of subordinated identities to a recognition of the productive possibilities of difference and border-crossing as such. I also describe how certain assumptions and tendencies in these approaches limit the usefulness of their analyses. In particular, the focus on language and textuality in these accounts obscures dimensions both of oppression and of liberatory possibilities in education; in addition, they are characterized by an affirmative impulse that tends to overlook the continuing importance of the negative—and indeed of class antagonisms—as a principle of radical politics. On the basis of this discussion, and in light of a consideration of the implications of globalization for cultural difference, I weigh the progressive possibilities of the notion of hybridity. This remains a crucial concept, although it must be rethought if it is to be adequate to a new historical moment. The notion of hybridity must be pushed even further, so that it does not rest in the mere difference of its elements, but combines and moves through them toward an original solidarity. The transnational scale of power and capital in the present can only be countered by a radical theory and practice that are capable of a similar scale and innovation.

Pedagogies of Difference

Affirming student identities necessarily involves interrogating the dominant discourses that marginalize different experiences, cultures, voices, and languages. The meaning of difference, however, as well as the conceptions of community and democracy that respond to it, can be understood in various ways. In this chapter, I discuss three important approaches to this problem in education: (1) If an assimilationist racism attempts to erase the distinct cultural identity of students of color and refuses to believe in their capacity to achieve, then to affirm their potential for excellence, and the value of their communities, is at the same time to challenge the dominant practices and attitudes that construct them as deficient. This is how the *culturally relevant* perspective describes the project of transformative pedagogy. (2) However, since subjugated voices themselves also reflect hegemonic values, and since the struggle is not only to free an already-achieved identity but also to resist closures forced upon the articulation of potential selves, student voices need to be interrogated as well as affirmed. The *critical postmodernist* perspective seeks to create a complicated pedagogical conversation in which affirmation and interrogation are not opposed but instead come together as complementary moments in the process of liberating possibility in the classroom and in society. (3) On the other hand, a pedagogy of *heteroglossia* exposes the disavowal by society (at large as well as in the classroom) of its own essential heterogeneity. From this perspective, the stifling of a potentially vibrant economy of diverse transactions within classroom discourse and culture more generally comes most of all at the expense of those whose cultural identities and practices are not deemed legitimate but also impoverishes communication and cognition for all students. In this model, the empowerment of subaltern meanings, languages, and identities involves the interruption of the dominant "scripts" that control the classroom conversation.

As all of these approaches suggest, valuing diverse experiences and understandings in schools is important in order to struggle against oppression and also to create a more enlivened and complex texture for classroom (and social) life. Official educational discourse, which centers "standard" English and white, middle-class culture, needs to be challenged but also unpacked and improvised against, so that the possibilities it forecloses can be restored by the irruption of different languages and knowledges. Difference (and democracy) in this sense are defined not only negatively, as freedom from domination, but positively as well, as the presence of a deep diversity. This vision challenges educators not just to counter authoritarianism but also to promote the proliferation of discourses. In what follows, I describe each of these educational paradigms in turn before moving to a consideration of problems in the senses of difference they propose, and how the relationship between difference and power might be more usefully conceptualized.

Culturally Relevant Pedagogy

"Culturally relevant" or "culturally responsive" pedagogy (see for example Gay 2000; Howard 2001; Ladson-Billings 1994; Villegas and Lucas 2002) seeks to recognize, affirm, and respond to the cultures of diverse students, in particular students of color, and to promote their achievement in educational systems that have been hostile to them. As it has been developed by Gloria Ladson-Billings (1992, 1994, 2001), the concept of *community* grounds this pedagogy, providing a framework in which the success of individual students can be supported. Community refers both to the society beyond schools and to relationships established in the classroom. However, the notion of community as the ground for pedagogy is based on a principle of difference. Culturally relevant pedagogy emphasizes the positive content of the cultural space that difference defines, against the subordination of it by the violence of the dominant. This idea is distinct from the mainstream liberal conception of difference as an equivalence between the multiple locations of a democratic pluralism. For culturally relevant pedagogy, success in teaching African American students, for instance, depends upon a recognition of the separateness and specificity of African American culture and community. This kind of difference resists being reincorporated into a simplistic democratic imaginary in which African American experience is just one of a number of interchangeable cultural locations. If racial and cultural difference has been a seam along which oppression of people of color has been organized in the United States, it also marks the outlines of positive and *distinct* histories of struggle, survival, and pride.

In this paradigm, educational progress and transformation depend less on recognizing others within a pluralistic framework, and more on the recovery and assertion of the strengths and values of a subaltern culture against ongoing domination. For individual students, this means urgently reconnecting with a grounding context that can promote their flourishing: "Culturally relevant teaching uses student culture in order to maintain it and to transcend the negative effects of the dominant culture" (Ladson-Billings 1994, 17). This sense of teaching as contextualizing and embedding students in community and culture is different from the idea that pedagogy should promote the acquisition by students of an abstract critical citizenship. Instead, in culturally relevant pedagogy, democratic education must dwell in the unique cultural identities that the differences of historical experience have defined. In the case of African American students, this means that it is essential to connect to the power of traditions of community and cooperation that have been centrally important in African American religion, education, and morality. However, rather than assuming any static attributes for students based on racial or cultural background, this approach emphasizes that effective teachers must take the time to get to know the contexts of their students' lives. This process of empowerment of children on the basis of their own cultural identities is not merely the precondition for a liberatory form of education but rather coincides with the creation of democratic conditions in school and society.

Culturally relevant pedagogy argues that in a world in which the crudest colonizing gestures are continually repeated (as for example when a teacher claims not to notice a student's race and in this way dismisses one of the most salient features of that student's identity), the real differences that power produces have to be made visible as a starting point for liberation. This does not mean affirming difference for its own sake but rather in order to challenge a process of erasure of the power of students to define themselves on their own terms. Culturally relevant pedagogy presents a challenge to dominant approaches to education and to all notions of teaching based on abstract models of citizenship. Instead, as Ladson-Billings demonstrates, pedagogy should attend to the concrete contexts, identities, and relationships that surround and define students.

Culturally relevant pedagogy bases the authenticity of teaching on a recognition of students' cultural realities and challenges the racism of deficit approaches with a commitment to the achievement of all students. It is important, however, for this approach to be sensitive to the dangers of closures that define culture and identity as decided contents. The assertion of culture must always be cautious, self-critical, and informed by an awareness of the presence in any identity of the Other. In addition, the articulation between community and achievement is a risky one, since the latter concept is overdetermined in schools by the logic of capital and in the present is constructed almost entirely as a matter of testing and credentialing. In this sense, culturally relevant pedagogy needs both a *positive moment* (the valuing of students and their capacities), and a *negative moment* (a refusal of the logic of current official educational meanings and processes).

Critical Postmodernist Pedagogy

While culturally relevant pedagogy focuses on the positive content of subaltern identities against their subordination by the dominant, critical postmodernist pedagogy (see for example Giroux 1992; Kincheloe and Steinberg 1997; McLaren 1997) emphasizes the instabilities, possibilities, and problems in all constructions of self, culture, and knowledge. This pedagogy points to the indeterminacy of social reality, as well as to the real political effects of different articulations of it. Henry Giroux, in particular, proposes a complex dialectic of difference and dialogue that informs a critical exploration and rearticulation by students of their own experiences and identities and of the world itself (see especially Giroux 1992; also 2000, 2001). In Giroux's account, difference is a crucial category, but he always means by this term the permeable sense of *border* rather than any primary separateness. In the context of the raced, gendered, and classed world that constructs student selves and the worlds they negotiate—whether in school, popular culture, or social life—a sensibility oriented to difference has two basic purposes. First, a primary task of pedagogy is to help students to investigate how they are differentially located in discursive systems, which are also systems of power, and to challenge the kinds of domination that result from the reproduction of these systems. Second, critical postmodernism uses the notion of difference in a productive sense to suggest the possibility of crossing boundaries. The differences marked by borders of experience, knowledge, and identity are invitations to students to risk an encounter with other selves and voices. In coming to understand otherness, as well as the contingency and fluidity of subjectivity, students are able to invent new and more democratic identities. This means promoting a proliferation of meanings and subjects against the unities of the old modernist paradigms of subject, student, and learning, according to which theorists attempted to characterize education in more or less general and universal terms. The modernist paradigms immobilize discourse within static categories, which for critical postmodernism is exactly how power functions in patriarchy, racism, and other forms of oppression.

However, this pedagogy does not just celebrate difference and contingency. Giroux's politics of "voice," for example, emphasizes both the importance of narrations of identity and the importance of questioning these narrations. Just as dominant institutions must be interrogated for the strategies of power they conceal, so too the voices of students need to be questioned, since they are also effects and expressions of power and contain their own limitations and contradictions. For example, male students oppressed on the basis of race or class may seek to demonstrate their resistance through masculinist and patriarchal assertions. The point of this pedagogy is to investigate how ideas and identities are located within circuits of power and knowledge, including in the voices of students. For this reason, interrogating identity does not mean simply reducing subjectivity to a multiplicity of fragments or traces but rather reconstructing it towards more informed and ethical kinds of agency. This position recognizes that students have multiple potential identities but also recognizes that students need a language in which to create a more just and equitable society. There is a link here with the liberal tradition within modernism, since critical postmodernist pedagogy also inculcates forms of solidarity based on the principles of justice, liberty, and equality. In the complex dialectic of this pedagogy, the invention of new identities takes place within a commitment to these fundamental ethical imperatives. On the other hand, in order to be realized, these principles themselves require the dismantling of the old categories of modernist thought that reinforce cultural, racial, and gender hierarchies in public life and in dominant approaches to teaching.

The philosophical position of critical postmodernism is defined on the one hand by foundational ethical commitments and on the other by an emphasis on the proliferation and interrogation of identities. These dual imperatives create an active tension that informs the notion of difference. Difference plays a central role in pedagogy, but left to itself it is potentially problematic. Difference must always be connected to universal ethical purposes: "Difference in this case cannot be seen as simply either a register of plurality or as a politics of assertion. Instead, it must be developed within practices in which differences can be affirmed *and* transformed in their articulation with historical and relational categories central to emancipatory forms of public life: democracy, citizenship, and public spheres" (Giroux 1992, 75). Here resistance against oppressive forms of discursive closure in pedagogy is explicitly tied to an appeal to modernist political principles. Giroux frames postmodernist criticism as a questioning of the modernist subject that nevertheless seeks to elaborate new forms of liberatory agency grounded in egalitarian commitments. The question is whether it is possible to appropriate, in a coherent way, elements of very different and frequently opposed theoretical movements. How does teaching undertake a general interrogation of the meanings that circulate through schools and students while also preserving a privileged set of understandings that remains to guide the whole process?

Pedagogy of Heteroglossia

In this approach, the sense of education as both collective and transcultural is the essential insight. Under this rubric, I have grouped together educational theorists for whom intersubjectivity is not just something to be cultivated as part of a progressive pedagogy but is rather the basic condition of education, constructing its limits and possibilities (e.g., Gutiérrez, Rymes, and Larson 1995; Lee and Smagorinsky 2000; Orellana 2001). Following Vygotsky (1978), this socioculturally oriented approach views cognition and literacy as constructed by the collaborative situations that mediate learning for participants. In this way, dominant educational designs, focused on individual and monolingual recitation, reinforce a sense of literacy as private submission to pedagogical authority; decentered and fluid communities of practice, on the other hand, allow for a flexibility of roles and (bilingual) communicative patterns that propose a sense of literacy as dynamic and constructed. There is an inherently political aspect to this work to the extent that it focuses attention on the complex social and cultural dynamics that learning begins from and the necessity of challenging educational practices that deny this complexity. With its emphasis on the multiplicity of languages and voices that students bring to the classroom, this paradigm challenges the monologism that has informed not only dominant ideologies but even some approaches on the educational left.

This approach displaces abstract and individualistic senses of culture and learning in favor of a view of these processes as essentially situated and social (Scribner 1984; New London Group 1996). In particular, in the work of Kris Gutiérrez and her collaborators, starting from Bakhtin (1981), the classroom is conceptualized as a space of heteroglossia in which different languages compete and collaborate to create an environment that variously promotes or impedes teaching and learning. This model focuses on the fundamental intersubjectivity and hybridity of educational processes, whether officially acknowledged or not. The problems of pedagogy are not solely the teacher's but the students' as well, since all share a responsibility for the texture of the conversation that takes place between them. This approach suggests a somewhat different way of thinking about agency in pedagogy, as compared with the culturally relevant and critical postmodernist paradigms. Whereas the latter elaborate theories of teaching designed to *develop* the capacity and power of students, from the perspective of Gutiérrez student agency is *already effective* in constructing the space of the classroom. The power of this subaltern student agency does not await the intervention of the teacher, or the "center" of power, to act, but instead already partly determines the discursive structure of the classroom. In this way, pedagogy is a "constructed text, a mosaic of the multiple texts of the participants" (Gutiérrez, Rymes, and Larson 1995, 468) in which the voices of students make their own autonomous contribution.

In schooling, processes of negotiation and resistance are mediated by differences of race, culture, and language that amplify the difference between the institutional locations of teacher and student. In the current system of public education, in which students of color confront an overwhelmingly white teacher force and in which the academic failure of children of color is increasingly normalized (Gutiérrez, Asato, and Baquedano-López 2000), the power relationship between teacher and student can reflect an almost colonial dynamic. Nevertheless, the pedagogy of heteroglossia identifies the moments that are richest in meaning and possibility not with those occasions when students simply refuse the language of the teacher but rather in instants of interruption that open the space for unscripted collaboration between teacher and student—that is, in the uncharted "third space" of genuine, if tenuous, dialogue. In this space, the teacher risks departing from the certainty of authoritative codes, and students risk an opening that makes their own counteridentities vulnerable to real engagement and to learning itself. Educational growth and meaning are supported not by repressing covert antagonisms, but rather by allowing them to emerge and to be confronted and transformed. In this process, identities partly fall away from themselves into a space that is "immanently hybrid, that is, polycontextual, multivoiced, and multiscripted. Thus, conflict, tension, and diversity are intrinsic to learning spaces. . . . By attending to the social, political, material, cognitive and linguistic conflict, we also have documented these tensions as potential sites of rupture, innovation, and change that lead to learning" (Gutiérrez, Baquedano-López, and Tejeda 1999, 287). At the same time, this process produces an organic conversation in the classroom—an ecology in which difference is a productive principle while being continually reconstructed by dialogue.

This paradigm analyzes how the teacher either offers his or her own voice up to engagement with students, or not, and suggests how educators can promote a richer and more democratic classroom conversation. This approach is especially helpful in thinking about how the teacher's voice can be relocated *within* the collaborative space of pedagogy, as *a* (central) element of it rather than reserving an inviolable authority insofar as it sets the terms of the educational experience. This paradigm helps to reveal those places in the classroom "text" where the teacher's voice becomes ungrounded and open to rearticulation by the narrations of students themselves. Becoming aware of these moments can help teachers toward a humility that places their own understandings *among* those of their students—not that these understandings are the same as their students', but rather that they coconstruct the meanings of pedagogy and curriculum. This is even more necessary in a world in which the "centers" of power pursue an ever more complete monopoly on authority and legitimacy, a tendency that deeply influences contemporary educational policy and practice.

However, this paradigm's understanding of educational life in terms of *textuality* tends to obscure the material backing that makes certain voices hegemonic while others are forced to struggle for a space in the margins. Likewise, the emphasis on collaboration and imbrication can distract attention from the ongoing importance within capitalist societies of antagonism and opposition. In this regard, the power of cooperation that is highlighted in this approach needs to be appropriated to *contest* the logic that marginalizes certain languages and identities in the first place.

Discourse and Power

It can be argued that the postmodern impulse to interrogate, explore, and reimagine identities is the very substance of democracy. From this perspective, difference does not simply mean the dispersion or fragmentation of structures and selves. Instead, it poses the problem of *articulation* (Laclau and Mouffe 1985): in a space of discursive openness, how should we choose to name and arrange the elements, to narrate ourselves and society in a provisional, strategic, and ethical manner? This sense of the strategic importance of discourse is central to the critical educational paradigms of difference that I have considered in this chapter. It underlies the emphasis on language and literacy in approaches that focus on cultural heterogeneity, such as the pedagogy of heteroglossia, and it crucially informs the effort in critical postmodernist pedagogy to create a dialogical space for the narration and interrogation of identity. Culturally relevant pedagogy, to the extent that it aims to combat deficit perspectives and practices, also depends very much on an understanding of the political power of articulation and counterarticulation.

However, there are problems with this emphasis on the politics of discourse, since as Terry Eagleton points out, "if our discourses are constitutive of our practices, then there would seem no enabling distance between the two in which . . . transformative labour could occur" (1991, 209). An understanding that focuses on discourse as the central terrain of politics sets aside the necessity of a critical reading that can find its way *between levels* of social life. How do teachers help students to understand racism, for example, as both a systematic organizing principle of society and as an encompassing discourse that conceals racist social structures and processes? Revealing the materiality of social reality does not mean reinforcing a naturalistic idea of the "objective," but rather exposing the difference between dominant representations of reality, which are the basis for institutional structures and practices, and the violence they hide, which is lived by those who are subjugated. This is the gap between representation and experience, which makes for a political dimensionality, namely the space of ideology. By contrast, educational pedagogies of difference tend to evoke a discursive surface without depth, in which narrations proliferate and recombine as students "cross borders of meaning, maps of knowledge, social relations and values" and in which pedagogy "decenters as it remaps" (Giroux 1992, 136). The notion of an *outside* that has some consistency through these decenterings is de-emphasized here, against which these mappings might be measured.

In this way, these pedagogies risk abandoning the space between surface and ground, which is the opening for criticism, politics, and opposition. Insisting on this space does not mean claiming access to fundamental causes, but rather points to the gap between representations and their material determinations. Arguing that voices need to be interrogated for the particular power effects produced in them is not quite the same as contextualizing students' voices against the background of persistent ideological structures and the systematic logic of social relationships that these structures work to hide. For instance, U.S. students may reproduce a crude nationalist rhetoric that they have absorbed from the broader media and culture. However, merely exposing the aggressiveness of this rhetoric and the way that it reflects the dominant discourse is insufficient; it is important, in addition, to explore with students the way nationalism functions as the basis for an imaginary solidarity that hides extreme inequalities in the United States and also how it acts as a means of winning consent for imperialist policies. Furthermore, what makes this a special problem in the context of education is that the space of ideological mediation very much resembles the space of learning. In both cases there is a movement between levels and a dialectical process of development that is more than the introduction of a difference (Freire 1997). There is something more at work in authentic teaching and learning than teachers making different narratives available to students and interrogating the limits of different voices. Of course, education is not just a simple passage from ignorance to understanding; nevertheless, teaching and learning do involve a process of incorporation and expansion in which difference constitutes a moment rather than the purpose or limit.

Cultural studies privileges discourse as the essential terrain of the political, while also understanding culture itself primarily in terms of signifying practices. This orientation has influenced critical education. In the cultural politics of the educational approaches considered in this chapter, language is the privileged domain for political and pedagogical intervention, and the different moments of its figuring of knowledge and power even come to stand for the historical: "Cultural studies is important to critical educators because it provides the grounds for making a number of issues central to a radical theory of schooling. First, it offers the basis for creating new forms of knowledge by making language constitutive of the conditions for producing meaning as part of the knowledge/power relationship. Knowledge and power are reconceptualized in this context by reasserting not merely the indeterminacy of language but also the historical and social construction of knowledge itself" (Giroux 1992, 164).

How pedagogy intervenes against a discursive backdrop that positions students in different ways in relation to authoritative formations of knowledge is a central problem. But history is more than language, and so the politics that confronts it must be more than discursive. Even given the importance of language, a cultural politics of discourse tends to ignore the shifts that can make other levels of social reality predominant. In the present, pedagogy has to confront a dramatic reconfiguring of the terrain of struggle in schooling from contests over the content of curriculum to the problem of the almost absolute *proceduralization* of teaching and learning. The contest is less about how different subject positions and epistemologies are valued or not and more over the ways that Black and brown bodies are controlled, tracked, and expelled by a system that is still structured (and now even more completely) in terms of what George Jackson (1970) once described as a militarist-racist-imperialist axis. Official educational regimes are now less concerned with situating students differentially within the "power-knowledge" couple and more concerned with a territorial management of them according to the principles of occupation and abandonment; what has been called in another context a new "right of the police" (Hardt and Negri 2000). Power in education is not simply a matter of forcing students to submit to the discursive violence of the dominant curriculum but is also manifested in extreme "zero tolerance" disciplinary rules, lockdowns, and drug searches (Lyons and Drew 2006), as well as high-stakes testing and promotion policies that result in pushing students out of school (McNeil 2005). In this context, transformative teaching has to be based on more than a politics of knowledge; it must be able to respond as well to aggressive processes of physical control and containment.

The educational paradigms I have discussed above outline a complex dialectic between solidarity and difference, identity and its interrogation, and ethical principles and the indeterminacy of politics. But it is also important that the proliferation of narrations and representations not obscure the interests that systematically organize the political terrain of teaching. Official "accountability" initiatives, for instance, systematically benefit upper-middle class white families not only because they reinforce the belief that their knowledge and culture are superior but also because they restrict access to college preparatory curriculum, higher educational opportunities, and professional occupations. They crucially reserve not only cultural hegemony but also access to wealth and power to the children of elites. In addition, we have to historicize any philosophy of oppositional pedagogy in relation to overall shifts in the institutional, social, and political contexts of education (Kincheloe 2007). The approaches I have considered here focus our attention on the goal of recreating schools as democratic spaces and on the construction of forms of solidarity that combine differences without denying them. However, pedagogical praxis toward that end must be sensitive to the specific kinds of closure that currently make democratic engagement impossible. Only from this vantage point can it propose a methodology capable of intervening in the historical present.

In addition, simply tracing the way that power positions identities within discourse and dialogue risks leaving the content of power unanalyzed. What does this power consist in that guarantees the teacher's centrality or the canon's hegemony? *Describing* the solipsism of the teacher's "script," which erases any response that does not reproduce the answer it anticipates, is not the same as *explaining* the power that enforces this script. We also have to ask what holds that center firm and guarantees its effectiveness. It is important to consider how the agency of students can be thought of as active and autonomous, already at work in the "text" of pedagogy and complexly interwoven with the teacher's own power, as Gutiérrez et al. show. And yet the notion of the classroom as an ecology should also focus our attention on the politics that comprehensively organizes this space. Beyond registering the fundamental heteroglossia of situations of teaching and learning, how can an oppositional educational politics challenge in practice the authority that stands in the way of articulating both students and teachers in a new way? What forces must be confronted in order to reconstruct teachers and students as asymmetrical coparticipants in learning?

If oppression were only the disavowal of heterogeneity, then ultimately it would fail, since the analyses of the theorists I have discussed here show the imbrication of all identities, including the most powerful ones, in each other. These theorists reveal the dishonesty of the authority and autonomy that are claimed by the powerful. Nevertheless, the authority of elites is in actual fact *backed up*—it is able to call on important resources to guarantee its hegemony. It is this backing that makes interruptions of it so difficult. In schools, the institutional forces that shore up authoritarian languages and practices are considerable. If there is a general heteroglossia that structures social space and is denied by the dominant culture, then beyond recognizing this fact we need to build radical hybridities that can challenge not only the monologism of the dominant but also the material force that permits this monologism to repeat itself indefinitely. The intertextuality of cultural and social life, including in education, is an important dimension of the oppositional identities that contemporary politics makes possible, but these new forms of opposition must be able to recognize their opponents. An effective resistance must value its own essential heterogeneity but must also not deny the fundamental antagonism that pits it against the exploiting and dominating classes. The accounts of difference considered above reveal the interdependence written into identities, but this recognition needs to become a starting rather than an ending point, and a basis, against oppression, for the realization of dialogue and democracy as material social conditions.

Hybridity: Problems and Possibilities

The flexible sense of difference that is expressed in contemporary critical educational approaches is grounded in a larger historical moment in which ethnic nationalist movements turned toward a more nuanced politics, emphasizing the constructed nature of social differences and the importance of relationships across boundaries, while retaining a sense of the centrality of uniquely different identities. Challenging essentialist accounts, theorists began to emphasize the specificity of cultural trajectories within broader shared histories (West 1988, 1993a); the deep imbrications of apparently separate cultures, even when opposed to each other in relations of colonialism (Said 1993); and the complexities of diasporic identities (Gilroy 1993). This more fluid notion of difference is related to a sense of culture as discourse, and discourse as a primary ground of the political (Hall 1986). In this way, culture is conceptualized as the product of historical choices and political strategies rather than as a pregiven and natural substratum of human being. In education, this theoretical turn has been associated broadly with a focus on literacy and language as political processes and with a form of critical pedagogy and multiculturalism that sees differences between students as important but also dynamic and socially constructed and that expresses its commitment to the valuing of difference within an understanding of the operation of power in schools and society (in addition to the work discussed in the first part of this chapter, see also Darder 1991; Nieto 1999; Sleeter 1993.) Feminist projects in education have likewise moved from an emphasis on essential gender differences to a focus on the necessary connection between teaching against patriarchy and teaching for radical change more generally (Weiler 1988), and even to a questioning of the very production of gendered identities within the context of schooling (Walkerdine 1998).

The investigation of cultural difference as historically constructed, as well as the emphasis on relationships between cultures and identities, has naturally been associated with an exploration of the process of crossing borders and with the notion of hybridity. In this vein, the problem of identity is immediately a matter of traversing boundaries, and retains an unsettledness that more stable senses of difference do not fully recognize. Gloria Anzaldúa's (1987) notion of "*mestizaje*," and Homi Bhabha's (1994) sense of "ambivalence," are examples of efforts to dwell in the difficult moment of hybridity, which refuses the simplifications of more familiar senses of culture, gender, and subjectivity. In this way, transgressing normative understandings and practices becomes a crucial progressive and critical project (hooks 1994). For some theorists, such as Judith Butler (1993), the exploration of the construction and limits of the normal means dismantling altogether any stable image of the social (and the real itself). For educators, this theoretical and political background has served as the foundation for the promotion of radical forms of hybrid language practices, as I discuss above in terms of the pedagogy of heteroglossia, which are themselves connected to complex senses of identity. Starting from some of the same theoretical insights, poststructuralist educational theorists have inveighed against pedagogies they deem inadequately sensitive to their own discursive closure (e.g., Lather 1998). However, the problem for educators, as for cultural workers generally, is whether an attention to hybridity and ambivalence serves to enrich a sense of solidarity in opposition to hegemony or whether it leads to the unraveling of useful critical categories and to the practices that they make possible.

Hybridity, like any other sign, is not fixed in its reference, uses, or political effects. As Stuart Hall (1986) demonstrated, while elements of discourse are constrained by social history, they can be variously appropriated, and it is important to intervene in this process. This is important in light of the fact that there is a more mainstream construction of hybridity, which is perhaps not so far after all from the exhilarating flights of the daring postmodernist ones. This is the sense of capitalism itself, particularly as represented in contemporary globalization, as a dynamically hybrid or at least hybridizing force. Globalization has been much celebrated as a process that brings previously disparate populations into relationship with each other and that opens a window onto a new global culture of the market and the new technologies. In a more nuanced version, this celebratory tendency does not deny the fact of conquest and colonialism but nevertheless praises the market for challenging traditionalisms, making possible new forms of global interconnectedness, and provoking a sophisticated cosmopolitanism (Appiah 2006). The export of Hollywood soap operas to Africa, from this perspective, should not be deplored as a form of cultural imperialism but instead recognized as a kind of cross-cultural communication; furthermore, such entertainment is understood in different ways by different populations. What is glossed over in these accounts is the culture of consumerism that arrives with globalization, as well as the material suffering produced by the disruption of economies and communities. These market-inspired visions of hybridity have a good deal in common, in terms of their enthusiasm for the recombinatory potential of culture, with many of the ostensibly subversive visions of hybridity offered by cultural theorists.

But cultural difference as hybridity can be articulated another way. The sense I want to suggest here is distinct from the subversions proposed by the poststructuralists and also from the critical multiculturalist and postmodernist perspectives that inform the educational paradigms I have discussed in this chapter. The vision of hybridity I propose is an oppositional one, one grounded not only in an impulse toward miscegenation that refuses the boundaries that separate meanings and identities but also in a revolutionary and utopian vision. In this vision, hybridity moves beyond the practice of border-crossing and in *combining* differences creates a new and collective identity and solidarity. One place to look for a prefiguration of this new hybridity is in the global movements against neoliberalism, privatization, and war, movements that are part of a complex transnational struggle against capital and its commanders. This oppositional notion of hybridity does not rest in the moment of difference but progresses from it to a new form of collective identity—a new and different species of social being. Here hybridity passes from a subversive sense of difference to a utopian vision of an oppositional commonality.[2] However, while the modernist vision of the socialist "New Man" was monumental, monolithic, and abstract, the image I suggest here is of a polyglot and territorialized circle of opposition that does not transcend its constituent elements in the forging of a universal figure but rather combines all of these elements and idioms into a new and extended culture.

Globalization conditions cultural difference in very specific ways. First of all, the aggressive penetration of capital worldwide has the effect of introducing increasingly common conditions of life to very disparate populations, even if this tendency is far from actually homogenizing different societies in any complete sense. But in addition, as globality itself as concept and horizon becomes increasingly familiar and comes to form part of popular schemas for understanding social reality, the differences located *within* this space begin to seem somewhat less crucial. Rather than experiencing the world as an infinite or unmappable universe, we increasingly experience it as a kind of locality. In this way, the apparently limitless regions and forms of life that constitute the earth are transformed into the smallness of a finite and embattled terrain. It is not so much that our world becomes the "local" to be referred to some vaster galactic "global" but rather that the very category of the global somehow shrinks, and we begin to understand, perhaps even at a metaphysical level, the fragility of totality itself.

In addition, at the level of globalization as political economy, it is important to recognize that capitalism binds people together in ways that are much more complex than often thought. Rethinking capitalism away from the simple paradigm of productivism, theorists have shown that its complex mediations should push us to think of it more properly in terms of a "capital system" (Mézsaros 1995) that reorganizes life across numerous dimensions, producing important contradictions within the domains of culture, nature, family, et cetera. In this way, various global social crises and the movements that respond to them, such as for instance ecological degradation and environmentalism, centrally participate in the working out of the dialectic of class struggle. Furthermore, much of what is usually considered to be external to capitalism as a mode of economic production—such as the varieties of cultural practice globally—may in fact be in deep relation to it, to the extent that these aspects of life become sites of important conflicts that express the antagonisms that define capital as a social logic. For example, Slavoj Žižek (2002) argues that the conflict between Islamic fundamentalism and the West, which is often described in terms of the struggle of cultural traditionalism against an invasive capitalist modernity, is in fact at once a conflict between two different expressions of global capitalist culture and two different expressions of fundamentalism (a religious and a free-market form respectively). In this sense, the capital system is a kind of disavowed universal within which "differences" strive (futilely) to project themselves as unique and authentic.

In this regard, critical educators need to explore hybridity and create a space for the expression and interrogation of difference, not as goals in themselves but rather in the service of producing a new and shared collective identity that brings together the different moments of the hybrid and links them in the production of something original. The task is not to rest in the incommensurability of different experiences but rather to propose a new solidarity on the basis of a constructed identity. In the age of globalization, hybridity needs to be conditioned by a sense of shared globality and should be rearticulated as the expression of that globality—the wholeness of its culture, still under construction. In this revolutionary sense of hybridity, a different "difference" is foregrounded: not the infinite gaps between local communities and identities but instead the difference between, on the one hand, a social logic of domination and exploitation and, on the other, a global movement and culture for liberation. This does not mean erasing the distinctness of different experiences, but instead gathering them together toward the wholeness that their complete collection proposes.

To the extent that multiculturalism translates histories of plunder and colonization into the liberal problematic of bias and to the extent that it ignores the class antagonisms that cut across individual differences, it colludes with capitalist culture and smoothes the way for the reproduction of many of the forms of social violence it condemns (Darder and Torres 2003). This perspective can end up trapping teachers and students into an individualistic economy of anxiety and obscuring the systemic processes that are busily organizing their futures. At the same time, however, a properly global consciousness means doing away with the epistemological privilege of the dominant—indeed of the First World in general—and rediscovering the true coevalness of global human communities (Mignolo 2004). Educators should appropriate from multiculturalism the emphasis on culture itself, the challenge to power, and the creativity inherent in bringing together disparate traditions and tendencies in the service of a revolutionary project of confronting oppression as a constitutive social logic. In addition to persistent discrimination, young people in the present confront the problems of insecure occupational and financial futures and also new forms of social anomie that produce not merely hopelessness but terror and chaos. Against these challenges, we should not back away from radical projects into the comfort of familiar discourses but instead plant new hope in spaces of objective possibility. In the present era of free-market fundamentalism and state terror, this space stretches beyond the boundaries of cultures, nations, and states: it is only a transnational oppositional movement that will be able to challenge capital along the entire length of its unfolding and at the level of its organizing principles. Therefore, young people must be able to imagine a global solidarity, and rethink their own identities on this terrain.

Conclusion

No paradigm of liberatory politics or education can be adequate that is not able to think through the problem of solidarity—to show how the complementary moments of interdependence and autonomy, for students and others, can be negotiated within radical praxis. The educational paradigms I have discussed in this chapter take on this challenge and provide powerful responses to it. Each describes, in its own way, how we can reinvent educational practices and purposes within the context of an acknowledgement of the importance of differences and difference itself. However, these paradigms tend to lose track of the oppositional relationship between the forms of solidarity they imagine and the oppressive forces that work to destroy them. In other words, in complexifying the process of dialogue across and within differences, they tend to overlook that difference, of a second order, between these enriched dialogical spaces and the larger social terrain on which they are constructed; this difference is defined by domination. In addition, in their preoccupation with the textuality of social life they tend to de-emphasize the materiality of power. The idea of educational politics as a sphere of competing, contradictory, and interpenetrating *languages* also limits the possibilities of education, since the broader logic within which these languages circulate and collide tends to become invisible in this perspective. This makes it difficult to propose forms of oppositional pedagogy as embodied praxis, since it appears that the job of teaching is to create a different *conversation* rather than to wrestle with a set of material obstacles to change.

In addition, in the age of globalization, difference and hybridity need to be rethought in terms of a new global oppositional identity. Hybridity in this context represents a dynamic and transnational movement against capital, and it is important for students to have access to the tools and information to participate in the construction of this movement. In this way, the idea of hybridity moves beyond the seriality of border-crossing toward the production of an original form of consciousness and agency. This sense of hybridity does not dissolve the complex shape and color of heteroglossia into the grey space of the economic or the sociological. A critical pedagogy dedicated to the idea of difference as revolutionary, and hybridity as the prefiguration of a new world, still draws life and energy from culture and still has faith in culture as earth shaking and history making. This vision transforms a commonplace of critical multiculturalism—that cultural differences are tied to different histories—from a principle for understanding already-given identities to a principle of contemporary oppositional praxis. To truly create and participate in culture, from this perspective, is to intervene in and make history. As students and teachers, activists and intellectuals, at the same time that we participate in producing a new politics, we also reconstruct our cultures, selves, and communities. In this process, difference is less the space to be preserved between discrete identities, and more the gap to be overcome between collective movement and the unprecedented form of life for which it strives.

CHAPTER 6

A Contemporary Philosophy of Praxis

In previous chapters I have focused on rethinking approaches to critical education based on contemporary political and cultural changes in schools and society. In this chapter, I am concerned with how these considerations can be extended to discussions of social movement and change more broadly. As John Dewey (1944) showed, almost all communication and sociality is inherently (or potentially) educational. In this way, political projects are more than sets of tasks to be carried out; they are also teaching projects and collective kinds of learning. A useful term in this context is *praxis*, which as reflective action that intervenes in a social context, necessarily transforming it, is a mode of being that cuts across many social domains, including education. A contemporary "philosophy of praxis"—to use the term of Antonio Gramsci (1971)—must be able to recognize the central dimensions of political and social reality in the present, including the contemporary organization of domination and exploitation, and propose a subject and process that might challenge this reality. A philosophy of praxis is more than a strategy for action, since it involves a theory of society. However, as Gramsci argues, it must be responsive to its historical moment. Critical philosophy has to take effect within society and must be able to discover contradictions that can be developed as points of reference for those who join the struggle. This broader philosophical and political project would then, in turn, constitute the framework for critical educational approaches in specific settings. In order to develop this praxis, however, it is necessary to engage in a critical reading, not merely of injustice but of power and capital, in an open-ended project of interpretation and dialogue.

In the present, any useful philosophy of praxis has to be responsive to processes of globalization. The emergence of a truly global economy, as well as a complex global culture, means that critics (and movements) have to think beyond the framework of the national and even the international (Sklair 2002). Familiar ideas for progressives, including class struggle, racism, and imperialism, have to be rethought in terms of the global. Marxist theory has been an important resource for thinking about globalization, and this tradition is a crucial organizing framework for my reflections in this chapter. And yet Marxism must be prepared as well to reconsider some of its basic assumptions in light of the actual experience of globality, as I will describe. It is a matter not only of tracing the complexities of a globalizing capitalism but also of proposing new forms of revolutionary or transformative consciousness, subjectivity, and action that can be responsive to a diversity of locations and struggles. At the same time, any adequate contemporary philosophy of praxis has to be sensitive to the difficulties of identification and consolidation that postmodernists (and the postmodern moment) have raised for political praxis. This is all the more urgent for proposals or theories of the global, since the possibility for the erasure of difference would presumably be even more dramatic at this scale.

The problem of describing a meaningful philosophy of praxis for the present can be usefully framed by two important challenges. First of all, power and capital have demonstrated a remarkable inventiveness in response to internal crises. This has led to an expansion of traditional forms of exploitation, as well as to new forms of economic and military aggression to shore up political and economic hegemony. David Harvey (2003) describes the widespread privatization, biopiracy, and neoimperialist practices associated with transnational capitalism as strategies that seek to respond to contemporary crises by releasing cheap inputs for capital to colonize and exploit. Anti-imperialist struggles, indigenous and antiprivatization movements, and others have emerged in opposition to this contemporary plunder. In the United States, struggles against school privatization, against the growth of the prison system, and against the wars in Iraq and Afghanistan are connected to this global movement against neoliberalism and neoimperialism. An important question, however, as Harvey points out, is how these movements can be connected, theoretically and practically, to each other, as well as to traditional working-class struggles against exploitation. What conception of global oppositional subject and practice can serve as the link between these various forms of resistance?

In what I believe is a second challenge closely connected to Harvey's, Chandra Mohanty (2003) argues for a "politics of location" that would seek to create forms of feminist solidarity that might cross national and cultural borders while avoiding the universalisms of the old left and of first-wave feminism as well. Mohanty suggests that any global political project must be able respond to the way that gender, sexuality, and culture are imbricated in processes of capitalist accumulation and must be able to trace the interconnections of these domains in different global spaces. For example, as she describes, patriarchal constructions of women's labor in many contexts as domestic leisure work provide cover for more intensive exploitation; both the labor and women's movements need to understand this interconnection between capital and patriarchy.[1] We can take from both Harvey and Mohanty a challenge to think, newly and carefully, a collective oppositional praxis in the present. In particular, they both call for theoretical and practical work focused on making links between different locations and struggles within the context of a sensitivity to the difficulty and specificity of such linkages in the age of globalization.

In light of this challenge, in this chapter I explore the usefulness and limitations of two different approaches to radical praxis in the present. I have chosen one paradigm that is quite recent, and another that is now rather venerable, in order to test how far the vocabulary that is already part of the history of critical theory and practice can take us. While it is important to think creatively in response to historical changes, it is also important to preserve as far as possible the analytical tools that are already available before abandoning them for others. The first (and older) of these critical paradigms is Paulo Freire's conceptual pair *oppressor* and *oppressed*, which framed his pedagogical and political theory throughout his career (see in particular Freire 1997, 1999; Freire and Macedo 1987). While this language is familiar to many activists who are not Freireans, his description of this antagonism is unique and suggests some important points of departure for praxis in the present. In particular, while in his account this contradiction is based to begin with on the dialectical antagonism of class struggle, it nonetheless opens our thinking about this struggle to a range of power relations in addition to class, as well as to their geographical organization; these characteristics are crucial in the context of globalization. The second paradigm I consider is the global oppositional subject—*multitude*—which Michael Hardt and Antonio Negri (2000, 2004) propose as the counterpart to their original paradigm of global sovereignty, which they call *empire*. *Multitude* is a unique and new idea, since it connects exploited and oppositional forces everywhere, and because it is not simply defined in opposition to power but rather as an autonomous organizing principle for society. I will argue that each of the formulations above responds, in its own way, to the challenge raised above to think creatively and synthetically about liberatory struggle in the present. And though each has its respective limitations as well, I will argue that the center of gravity for theorizing power and resistance in the present should remain somewhat more within the first framework than within the second.

Finally, on the basis of this investigation and comparison, I will propose my own framework for thinking about oppositional movement in the context of globalization. This paradigm, while imagined on the global scale of Hardt and Negri's *multitude*, resists the deterritorialization that is basic to the latter. In reframing the notion of class struggle, it attempts to stay true to the dialectical determination of Freire's oppressor-oppressed couple while nevertheless enlarging and complexifying it. I believe that my proposal here constitutes a return to a Marxian conception of humanism while at the same time opening this project to a broader planetary commitment. In making territorialization and situatedness, *on a global scale*, the very content of a revolutionary class identification, it suggests an original answer to the challenges posed by Mohanty and Harvey that I have described above. This new framework also then suggests some important directions for organization and pedagogy, and I will close with a discussion of these.

Dialectic of Domination: Oppressor and Oppressed

Contemporary capitalism is characterized by hypermobility, supranationalism, and disregard for all previous boundaries and obstacles to its expansion. These characteristics generate a series of new social catastrophes while also potentially setting in motion a new set of oppositional actors. Seemingly progressive movements and societies come under its sway, as for example in South Africa, where the revolutionary African National Congress, which led the struggle against apartheid, now leads the way in innovative programs to squeeze working people and the poor (Bond 2006; Desai 2002). Seemingly successful developing economies are suddenly brought low through crisis, economic restructuring, and devaluation, as in Latin America, as transnational capitalism exploits temporary crises to suck whole societies dry (Harvey 2003). And fantastic global migrations are set in motion by the demands of the global North for cheaper labor, not just abroad, but at home. The current firestorm over illegal immigration in the United States indicates the shock waves of one instance of this larger moment, as anxieties over an unstable and globalized world are displaced onto demonized minorities in a futile attempt to preserve some sense of stasis and nation (Appadurai 2006). Globally, including in North America, public sectors and services, not least education, become the stalking ground for privatizers and free-marketeers. In the context of these dramatic upheavals, new movements of the dispossessed appear and take on a variety of forms, from state-centered proto-socialism (e.g., Chavismo) to the explosion of a multiplicity of NGOs worldwide, to new forms of labor organizing focused on casualized and unemployed workers, to independent social movements committed to the creation of autonomous politics or even territories (e.g., the Zapatistas). In this proliferation of movements, what political understanding and project can serve as a compass? What oppositional identifications can both empower people and also expose the links between different instances of exploitation and resistance?

One possible model for orienting resistance to power and capital in the present is Paulo Freire's problematic of oppressor and oppressed, which reframes the traditional language of class struggle on the left. The durability of this paradigm, familiar from educational and social struggles in the last several decades, is instructive. The innovativeness and flexibility of Freire's analysis of power makes it potentially useful for contemporary social movements, as I will describe, even if its application to present-day struggles means developing and extending it in new ways. The antagonism between oppressor and oppressed is quite different from the contradiction between the bourgeoisie and proletariat in Marxism. At the same time, however, the logical framework of traditional class analysis is important to Freire's theory, even if for the most part he avoids the orthodox categorizations. The language of *oppressor* and *oppressed* is designed to call attention to the lived, experiential realities of the day-to-day reproduction of domination in the lives of people. This means that a liberatory program that seeks to counter oppression must center the humanity of people through the process of *humanization*—indeed, that class struggle is in fact a kind of rediscovery of this humanity. The historical dialectic of class struggle becomes, as humanization, the "people's vocation," which, even if it "is constantly negated . . . is affirmed by that very negation" (Freire 1997, 25). The language of oppressor and oppressed highlights this relational dimension of class and inscribes in the vocabulary of critical pedagogy not merely a formal topography but the active transitivity of political relationships. Nevertheless, this attention to the existential complexity of power never contradicts Freire's insistence on class relationships as constitutive of the structure of society.

The framework of oppressor and oppressed also makes it possible to recognize the importance of a range of different power relations, as well as the ways in which they contribute to constituting the class contradiction itself. Relations of power connected to culture are highlighted in Freire's framework, and he describes the cultural marginalization and pathologization of the oppressed as a central modality of oppression (1997, 54–58). For the oppressed, the process of liberation is a process of recovery of themselves as authentic producers of culture. For example, Freire talks explicitly in many places of linguistic relations of power and focuses on literacy as an arena of class relations and liberatory possibilities. Furthermore, the framework of oppressor and oppressed makes it possible to organically connect class and cultural contradictions to those of racism and sexism, even if Freire himself attends somewhat less to these possibilities. In this regard, there is an important similarity between Freire's rearticulation of the Marxist analysis of class relations and the "stretched dialectic" of Frantz Fanon (1963), who in his own way adapts this class framework to the situation of colonialism. For Fanon, the dialectic of class struggle has to be rearticulated to recognize the centrality of racism not merely as a fact of colonialism but as a central structuring principle. Freire's analysis dislocates the traditional dialectic in a similar fashion by maintaining the central antagonism while extending the range of reference of each of its terms.

Furthermore, rather than positing the dialectic of class antagonism in abstract and formal terms, Freire's framework suggests that this antagonism is always inflected by territoriality and regionality. Oppression has to do with the spatial organization of power and its projection across spheres of influence. This sense of the spatiality of oppression gives his account a special relevance in the context of globalization. In one striking description of what Freire calls *banking education*, he writes: "The student records, memorizes and repeats . . . without perceiving . . . the true significance of 'capital' in the affirmation 'the capital of Pará is Belém'" (1997, 52). This sentence expresses the basic idea of the stripping of historical processes and political horizons from the content of the curriculum, but in addition it suggests the reproduction of the *geography* of power—the creation of elite center and margin, of capital and province. Similarly, Freire's lifelong reflections on Brazil and his home state of Pernambuco (see Freire 1999) indicate this preoccupation with the politics of territoriality and its reduction of the people of the rural Brazilian Northeast, from the perspective of power, to mere effects. The class contradiction, then, is mediated by these terrains—the material topographies of landscape, culture, and political economy. Freire's methodology of *problem-posing education*, in this context, is a kind of *travel*—a transportation of consciousness across the distances of disempowerment, a confrontation of the capital by the provinces, and a recontextualization of those places and meanings that have been centered within the larger framework of their relationship to the margins. Examples of this traversal of distance can be literally seen in the demands for rights and recognition in European societies by immigrants from former colonies, or in the confrontation—through fences, security zones, and police barricades—of the global elite by protestors from all over the world at the meetings of international financial institutions.

Freire's consideration of global relations of dependency and the immobilization of Third World societies in the disabling relationships of neocolonialism shares much with the work of anticolonial, postcolonial, and world systems theorists. But in its sensitivity to context and its mapping of social relationships in concentric circles from local to regional, Freire's work has an originality that has contributed to its usefulness in diverse global locations. These principles are best expressed in his concept of the *generative theme*, which names a central problem or contradiction in a given social context. According to Freire, generative themes need to be isolated on a series of levels, from the continental to the national to the regional to the local (1997, 84). At each of these levels, experience is characterized by particular *limit-situations* that it is the job of the educator to investigate prior to embarking on any project.[2] This refusal to specify the terms of the problem beforehand, as in his work in Africa, clearly challenges the universalistic impulse within the orthodox Marxian problematic of class struggle. Instead of importing ready-made schemes to a literacy project in Guinea-Bissau, Freire and his team from the Institute for Cultural Action listened to and studied the local context in collaboration with participants before conceptualizing and undertaking any projects (Freire 1978; Gadotti 1994). In this way, rather than seeking to reattach lived limit-situations, through skilful displacements and condensations, to a central and pregiven contradiction, the framework of oppressor and oppressed allows for more flexibility, since the priority of local contradictions can then organize analysis and lead the way.

The wide applicability of Freire's work also points up the centrality, within countries of the global North as well as the South, of the territorial logics of *marginalization* that he analyzes. Above all, the forms of dependency that he critiques are exemplified in the discipline governing so-called developing societies, which links cultural and economic violence against the oppressed to the global map of crises of capitalist accumulation. Freire's perspective on class is important in the context of a neoliberal world order characterized by extreme austerity measures imposed on the global South, as well as the pervasive and spectacular privatization of every resource and dimension of life. The logic of neoliberalism represents the return of "banking education" with a vengeance. In Freire's words: "The teacher disciplines and the students are disciplined; the teacher chooses and enforces his choice, and the students comply; the teacher acts and the students have the illusion of acting through the action of the teacher" (1997, 54). Neoliberalism represents the globalization of the most extreme version of this form of regulation of society, subjects, and knowledge and constructs the poor of the world as essentially lazy and deviant, in need of the iron discipline and expert knowledge of the international financial institutions and their local proxies. The demonization of the oppressed that Freire analyzes as central to elite narratives is alive and well in this logic. For example, Ashwin Desai describes the wholesale eviction, by the state, of poor people from their homes in the new and neoliberalized South Africa, in which the ruling African National Congress "presumes that poverty-stricken township dwellers are social deviants by virtue of their degraded circumstances" (2002, 54). This is a brutal form of what Freire called *assistencialism*, in which the condition of receiving aid is entrapment in dependency and surrender to a neocolonial logic that forces whole societies onto a path of development that is supposed to emulate the vicious capitalist culture of the north. In this universe of capitalist globalization, as class struggle is increasingly mediated by North-South relations and by relations between elite and subaltern populations, Freire's flexible and open-ended problematic of oppressor and oppressed remains extremely useful.

Harvey argues, as I have noted, that in the context of a worldwide capitalist crisis, alternative forms of accumulation have become important. As I describe in Chapter 4, he shows that what Marx called *primitive accumulation*—the original expropriations that characterized the enclosure of common lands in Europe and the plunder and pillage of imperialism—has remained important throughout the history of capitalism (Harvey 2003). In the present, in new forms, it is a key mode of accumulation within the regime of neoliberalism. In "accumulation by dispossession," as Harvey calls the contemporary form of primitive accumulation, rather than societies or sectors being smoothly incorporated into the regular functioning of mature capitalism through expanded reproduction, assets are taken through coercion in order to make possible cheaper inputs and thus more possibilities for the profitable investment of overaccumulated capital. The privatization of communally held resources globally and the seizure through imperialist practices of land and energy sources are examples of this form of capitalist accumulation. When people are forced to pay for services such as utilities that were previously provided free by the state, or when indigenous knowledges or even organisms are patented by multinationals, capital takes advantage of a form of direct, coercive expropriation that sucks dry an accumulated store of social value. Harvey suggests that what an oppositional class politics needs in the present is a way to link struggles against these dispossessions with more traditional struggles around capitalist reproduction; that is, to bridge movements against imperialist invasions and expropriations on the one hand and working-class and trade union struggles on the other.

The conceptual framework in Freire of oppressor and oppressed provides an important starting point for this challenge from Harvey. Freire's sensitivity to the geography of power can allow us to see the ways in which class is always territorially articulated. To struggle over the process of *exploitation* means first of all to understand the (spatial) logic of *marginalization*, both locally and in the context of global relationships. Freire's work begins to supply us with a vocabulary for understanding, in existential and ontological terms, the kinds of dispossession that characterize neoliberalism. Marx (1973) described capitalism as a system in which the inherent creativity of workers is alienated from them and becomes the "objectivity of an alien subjectivity" in being controlled by the capitalist; that is, labor power, a basic human potential, is appropriated as a commodity itself. In Freire's account in *Pedagogy of the Oppressed* (1997), this alienation is even more total, as the very possibility of authentic human subjectivity and spirit is pillaged by the powerful, and as the oppressed are reduced to a state of petrification. Indeed, the powerful come to monopolize historical agency and humanness itself. In short, the expropriation that Freire describes is different from the mere alienation of labor power in the context of capitalist reproduction. In the relation between oppressor and oppressed there is an additional "accumulation," if it can be called that, which is that of spiritual plunder, pathologization, and immobilization. This plunder expresses a state of bondage that is closer to Hegel's dialectic of master and slave than to the traditional Marxist account of the relationship of exploitation in mature capitalism.

This synthesis in Freire of the material and the existential can begin to suggest links between identifications forged around struggles against "accumulation by dispossession" and those that are the result of struggles over the process of capitalist reproduction. His flexible understanding of oppression can potentially be a crucial bridge between the ends and self-understandings of working-class struggles on the one hand and movements against cultural and economic imperialism on the other. While scholars have rightfully pointed out the important difference between the desire of Native peoples for political and cultural autonomy and the universalizing impulses of critical theory (Grande 2000), globalization reveals the deep links between annexation and capitalist exploitation. Thus, the movement by Native Americans in the United States and Canada to reclaim their lands and cultures has much in common with resistance against mega-development projects in India and the movements of Mexican *campesinos* and *guerrilleros* against neoliberalism, to the extent that each of these instances represents a refusal of the right of capital to subjugate, plunder, and absorb in the service of ever-widening and ever-greater accumulation.[3] Likewise, even though the process of "structural adjustment" that the less powerful countries have faced has not yet been visited to the same extent on the societies of the powerful countries in the global North, there are signs of it even there in the corporatization of education and other sectors. The convergence of diverse struggles globally requires a vocabulary that can help to organize these movements, and Freire's flexible problematic of oppressor and oppressed contributes an essential, if still incomplete, framework for rethinking class struggle on a global scale.

Swarm of the Global: Multitude

Globalization manifests itself not only in terms of a ubiquitous and predatory capitalism but also as an increasing interconnectedness of societies, economies, and populations. Of course, these are often one and the same process, so that the substance of this interconnectedness is in large part the condition of a common subjugation to a global regime of expropriation. Nevertheless, both within and against this regime, new forms and possibilities of collaboration and resistance are emerging. This is due not only to technological innovations in communications and the Internet but also to the movement of peoples and cultures. Is this the beginning of a new internationalism—or something newer still, the emergence of a global subject beyond and outside of the nation? And what forms of organization and action will this new subject produce? One of the most intriguing sets of responses to these questions is offered by Michael Hardt and Antonio Negri. In *Empire* (2000) and *Multitude* (2004), these thinkers develop several central tendencies within Marxian thought in a decidedly unorthodox and original direction, and propose both a new conception of sovereignty and a new subject of democratic opposition. Their work offers a number of very creative solutions to the challenges outlined at the beginning of this chapter. Their proposals are also marred by some important flaws and difficulties, which it is equally important to be attentive to in the course of developing a contemporary philosophy of praxis.[4]

The crucial concept from their work, for my purposes here, is *multitude*, which is the name both for a new mode of production of social life and for a new class and political project that links working people, the poor, and creative humanity generally in a struggle against "empire." *Empire* is a polycentric and immediately global form of rule, which joins the process of accumulation with a new networked form of political sovereignty in a single force. Likewise, the *multitude* is a supranational figure that links the innumerable singular instances of social production to a common project of global democracy. The World Social Forum and the global antiwar movement are examples of the multitude in action, but so are antiprivatization movements from Bolivia to South Africa. As they see it, even revolts responding to apparently locally embedded issues—from the Los Angeles uprisings to the Palestinian *intifadas*—reply subterraneously to each other as part of a single instantaneous articulation of global democratic aspiration (2000, 52–59). In this sense, Hardt and Negri both extend and exceed the Marxist project of articulating an international subject of oppositional praxis. The multitude is more than international—rather than overwhelming the state, it unravels the category of the nation altogether in instantaneously linking labor and creativity in global networks and in rearticulating opposition beyond the struggle over state power and instead locating it in struggles around basic forms of life. If the development of Freire's notion of oppressor and oppressed opens up possibilities for combining a range of movements and struggles, the project of multitude is explicitly motivated by this imperative. The problem posed by Harvey of linking opposition to new forms of dispossession with established forms of struggle around capitalist reproduction is central to the idea of multitude. Hardt and Negri seek to bring struggles together and to reveal the fundamental commonality between them as struggles within and against empire.

In order to do this, they discover the common basis of exploitation, annexation, and war in the idea of *biopower*, which operates within the realm of economic production narrowly conceived but also across all domains of human sociality. In their own articulation of this concept, they marry Foucault's complex figuring of power to the Marxian conception of capital. The regime of biopower is a "form of rule aimed not only at controlling the population but producing and reproducing all aspects of social life" (Hardt and Negri 2004, 13). Power and capital, as empire and biopower, come together at once in the increasingly total organization of everyday life by a new kind of sovereignty and in a more profound capturing of surplus value in the economic sphere. Surplus value in their account is thought of not only as a portion of the abstract value produced in a working day (as in Marx) but also as the very forms and networks of communication and cooperation that the multitude invents in its incessant and essential creativity. For example, even if the spectacular migrations of people and culture across the globe, which create new webs of association and complex global identities, are driven by the demands of capital, they nevertheless also represent an autonomous energy and desire of the multitude, a resource which is then exploited and organized by empire. From this perspective, the disciplining of economic production and the reconfiguration of political space on the basis of perpetual war are ways that power as top-down globalization seeks to capture and command the unlimited potentiality of social life.

The "becoming common of labor" that defines the multitude in Hardt and Negri's account means the production of new and shared forms of communication, cooperation, and being—new forms of society itself (2004, 103–15). In this sense, empire is dependent upon multitude not merely for the tribute that is taken from it to feed the machine but also for the creative forms of sociality that it invents and that are then appropriated to construct the network of empire. This represents the culmination of a long tradition of autonomist Marxist theorizing that has shifted the historical priority, in its account of the development of capitalism, from the forms of capital to the struggles of the working class (Holloway 2002). In the analysis of Hardt and Negri, contemporary work is characterized most importantly by the rise of forms of "immaterial labor" in the context of the information economy and the service sector that require a disciplining and organization of affects, the body, and the natural creativity of communication. (One can think here of the emotional work of telemarketers and flight attendants but also, more abstractly, of the innovations of software designers.) While the subordination of these economies to the processes of accumulation and reproduction of empire represents a new and more total form of exploitation, it also indicates the tremendous power of communication and social cooperation within the multitude. The immaterial labor of the multitude, redirected toward the emancipatory project of global democracy, releases hitherto unimagined possibilities, as the tremendous power of networks are oriented toward creativity and transformation. The global antiwar and antiglobalization (or "alterglobalization") movements are the most visible evidence of this emancipatory power, but it has to be remembered that new technologies, new social habits, and global cultures are all ultimately the expression of the inherent creativity of the multitude. If we are attentive to this evidence, they argue, then our hope and faith in revolution (reconceptualized from the taking of state power to biopolitical transformation and the materialization of global democracy) is reassured and sees the emergent future everywhere.

This ingenious set of conceptualizations also proposes a solution to the central problem posed by much postmodernist political and cultural theory: how to imagine an oppositional subject that does not do violence to social difference in constructing its hegemony. Hardt and Negri suggest that the multitude is a project rather than a unity—the name for the opening to democracy that is evident in contemporary global politics (2004, 219–27). As such, the multitude does not organize particulars into a universal category or enforce an assimilative identification that submerges the differences between movements. Instead, multitude is the name for the commonality that is created rather than essential in global sociality—a commonality that can also be thought of as the coming together of singularities into a swarm or network. In this network, no node is subordinated to an authoritative center; rather, each retains its specificity, and the larger commonality is glimpsed only as an abstract potentiality that does not violate the singularity of each node.

Hardt and Negri believe in forms of organization that emerge out of the activity of the multitude, and they reject the centralization of authority that is characteristic of the state, as well as of many revolutionary movements: "In contrast to the transcendental model that poses a unitary sovereign subject standing above society, biopolitical social organization begins to appear absolutely immanent, where all the elements interact on the same plane. In such an immanent model, in other words, instead of an external authority imposing order on society from above, the various elements present in society are able collaboratively to organize society themselves" (2004, 337).

In this sense, it is possible to see in the multitude a more expansive expression of Rosa Luxemburg's (2004) conception of the "mass ego" of the working class, which alone can truly be the subject of revolution and truly learn its methodology through experience. In her account, as in Hardt and Negri's, social movement is the organic product of situated learning in the context of collective communication and action. However, the concept of multitude departs from the assumption in Marxism, as well as in critical pedagogy, that opposition depends upon a process of conscientization that is guided by a teacher or leader, even if that leader's authority is dialogical and democratic. The model of learning proposed in the concept of the multitude is that of an orchestra without a conductor—a polycentric intelligence in constant communication with itself. In their conception, praxis as education is a kind of self-teaching of the multitude that is organized by intelligence and experimentalism but dispenses with the facilitator.

Hardt and Negri's account is energetic and inspirational. However, part of its effect comes from its distance from the very instances of exploitation and resistance that it aims to illuminate. The concept of multitude is supposed to preserve the differences between singular instances against any assimilation to a pregiven unity. However, their referral of all moments of struggle (e.g., the Zapatistas, the antiwar movement, etc.) to the grand abstraction of a single figure ends up partly deterritorializing them. At this higher level of abstraction, where the poor are found to propose a special ontological plenitude and a "swarm" intelligence links the innovations of open-source computer programmers to the tactics of street protesters, things look sunny and bright. (Indeed, their inveterate optimism sometimes recalls the tone of outright globalization boosters such as Thomas Friedman [2005], for whom the penetration of global capitalism into every corner and its reorganization of social life is cause for unambiguous celebration.) But a theory of praxis must be sensitive to the total experience of struggle, including its moments of loss, failure, and despair. As Holloway (2002) suggests, this means holding to a dialectical understanding of revolution that starts from the negation of oppression rather than the simple positing of a complete and undamaged autonomous agency. Hardt and Negri claim a formal humanism that critiques the philosophy of transcendence. But what they neglect is the recognition of human suffering as the starting point for politics—and the possibility of a notion of the "human" that is inherently linked to this history. If, at a higher level of abstraction, processes of privatization, global war, and the disciplining of labor indicate not only the emergence of empire but also, in spite of themselves, the unstoppable power of the multitude, this cannot provide much reassurance to those who are injured by these processes. And to the extent that the notion of multitude (like "globalization" itself) too eagerly participates in the destruction of social and political identities that have given people strength and hope in the face of oppression, then it may do more harm than good.

In addition, their proposal does not work out practically the difficult project of solidarity. To undertake this is necessarily to name names; for instance, constructing an effective solidarity means recognizing how the experiences of women who are exploited and injured by imperialism are different from the experiences of women (and men) who are not, as Chandra Mohanty (2003) shows. It means confronting whiteness, colonialism, and the racial economy of capitalism. It means investigating how patriarchy is used by capital and how racism is reproduced by exploitation. A respect for the difficulty of the political and for the necessity of engagement in a dialogue toward effective and liberating forms of solidarity is the ground of the convergence of trade union and environmental protests against neoliberalism, for instance, and underlies as well the prospect of any real politics of "race traitorism" by whites in the context of antiracism. The notion of multitude opens up a theoretical horizon for thinking resistance beyond the framework of the nation and outside of the familiar dialectic of the particular and universal, but it does so at the expense of political understandings that have been built in the course of historical struggles and whose usefulness may not in fact have entirely expired at the turn of the millennium (Aronowitz 2000). In their effort to capture the essential turn of the epoch and to write both the *Capital* and the *Manifesto* of the age of empire, Hardt and Negri assume rhetorically what Said described as "an invisible point of super-objective perspective" (1993, 167) and partly lose touch with the always-situated spirit of revolution. Their account is evocative, but the prospect toward which they gesture should be approached from a different direction and with a different political-ethical compass.

Class, Solidarity, and the Politics of the Terran

How can we learn from the accounts I have discussed and also avoid their impasses? What conception of oppositional praxis can respond to the questions put forward at the beginning of this chapter and still maintain a connection to practical struggles? The centering of the contradiction between capital and labor seems too constrained as an analytic to encompass the variety of social struggles in the present, and for many appears to force us toward a univocal reading of the social world. And yet the violence of state and capital are continually reproduced, and not to attempt a coherent account of their structure and organization would be an irresponsible abdication. On the one hand, it seems presumptuous to pretend to name and know the identities that organize social struggle globally and that constitute the framework for democratic struggle. Hardt and Negri's empire and multitude have a blockbuster-ish monumentality to them that can end up sounding like an effort at branding rather than a way to be attentive to power's various forms. And yet the reality of localities and individual lives is increasingly permeated by global forces that appear as such even to these individuals; this creates both a responsibility and an opening to which radical praxis must be responsive. In this context, what form should a radical reading-through of power take, and what kinds of syntheses should it aim for?

Terrains of Struggle

It is important for critical theorists to preserve the tension between objective and subjective dialectics of struggle, between dispassionate analysis on the one hand and outrage on the other. The progressivism and dynamism that characterize capitalism as a historical force should never be considered a compensation for its brutalities; conversely, anger at these brutalities needs to be informed by an attention to the objective logic that produces them. A materialist analysis understands that life is characterized by an irreducible antagonism between social forces. But if this idea is not to end up as a simple formalism, it needs to be filled in and illuminated by lived experience. Even the *concept* of struggle, after all, is something more than the formal idea of antagonism; its very idea points to the experiential. To properly understand struggle as the framework for politics means to be affected by it and to be made a partisan within it. If this is true in the context of globalization, it means that the forms of outrage, suffering, and resistance that globalization gives rise to must be in large part the basis for a correct analysis of its meaning. In this way, there is an ethical core to political understanding that cannot be rendered superfluous after the fact once we arrive at the right analysis. This ethical-political conception is captured in Freire's idea of humanization, which is connected both to historical logic and to an ethical imperative for all people as human beings. The material of human theory is human experience, and the historical and material solidarities that have bound together human bodies in struggle are the ground of politics.

In light of this notion of the political as based in lived histories of struggle, we should challenge Hardt and Negri's impulse toward an easy optimism in relation to globalization. In their account, there is no room for what they consider to be a reactionary nostalgia for the preglobal. All new creative and oppositional possibilities take place on the new terrain of the global, which is itself produced out of the dialectic of empire. From this perspective, actually existing struggles against capitalist globalization—whether they are struggles for indigenous rights or old-fashioned statist interventions—appear at least somewhat backward. And on the other hand, the violent deterritorializations that characterize neoliberalism—the uprooting of populations, the devastation of ecosystems, the assault on traditional identities—appear as ultimately progressive according to the logic of these authors (since in more fully materializing empire, they create an opening for multitude). But class struggle has always involved fighting against the erasure of bodies, identities, cultures, and experiences. To argue for the value and defense of these is not to argue that they should not be transformed in the course of opposition. Instead, it is to enlarge the set of experiences that come to define the content of an oppositional class, and to suggest a new project—the confrontation, negotiation, and imbrication of these experiences within a new class consciousness.

In the first place, this means arguing for a form of oppositional solidarity that persistently foregrounds different instances of oppression and recognizes itself in the range of oppositional formations that rise up against domination. This form of solidarity would constantly collect its different instances and members, as of an extended family, while resisting their assimilation to any single formulation. Mohanty writes: "Movement among cultures, languages, and complex configurations of meaning and power have always been the territory of the colonized . . . It is this process, this reterritorialization through struggle, that allows me a paradoxical continuity of self" (2003, 122). This means a constant attention to the actual experiences of colonization, patriarchy, and class oppression and to their complex links and syntheses. A loving and militant solidarity refuses not only the violations of power and capital but also the tendency of even a left "politics of transcendence" to abstract from these struggles in the name of an arbitrary unity.

In addition, however, this solidarity should be pushed a step further, so that an oppositional class project can be built from it that is comprised of the aspirations of the colonized, oppressed, and exploited and that is at the same time a space for the formation of a new identity. It is important to emphasize that this idea of class is not that of an objectivity given by the movement of capitalism or globalization itself but is rather the conscious political production of participants themselves. This represents a return, in a profound way, to Gramsci's notion of hegemony as a thoroughly political and educational project—the production of a bloc that can act effectively in history and yet whose purpose and meaning is not given a priori. At the same time, in linking diverse struggles and participants, the bonds that would hold this insurgent hegemony together would have to be stronger than those simply of political strategy. These bonds would necessarily be created as well from the irreducible solidarity and love of living beings for each other—the recognition of an essential kindredness and affiliation of consciousnesses for each other as members of a human and planetary family.

Reinventing Class

One of the central problems in the present is how to understand the coming together of politics and economics, of power and capital, in a unified figure of rule and social reproduction. The seizure of common resources can no longer be clearly distinguished from the regular production of surplus value; the ideological production of consent can no longer be clearly distinguished from the real sociality it is supposed to represent; the expulsion of bodies from the social (e.g., in the prison-industrial complex) and the new forms of exploitation of their labor power and subjectivity can no longer be clearly distinguished from the incorporation of bodies (in the workplace, schools, etc.) that is at the same time also now a process of organized abandonment. The notion of empire responds to this reality very creatively in bringing together these contradictory modalities of power and capital within an attention to the persistence of contemporary global crises that require a continual process of biopolitical management. However, the idea of empire too quickly subsumes social contradictions into a unified category and too quickly abandons the critique of capitalism as a framework for making sense of power and resistance. It would be better to avoid this impulse toward a grand solution and instead to complexify our sense of the dialectic itself—i.e., to rearticulate the vocabulary we have in order to be responsive to the present. This is not only because it is better theoretical practice to preserve concepts if possible (e.g., *class, power, capital, imperialism*, etc.) than to dispense with them, but also because these categories contain within them echoes of struggle that it is dangerous to forget.

What would it mean then to maintain the concept of *class* as central while rethinking it at the same time? First of all, to emphasize class means to emphasize that struggle is collective and that collectives in struggle are something more than the set of individuals that comprise them. In addition, to retain the language of class is also to retain an emphasis on social contradiction or antagonism, since a class only exists in opposition to another class within the overarching framework of class struggle. However, as I have described above, it is necessary in the present to abandon the idea that class consciousness only fills in the previously existing placeholder of a class-in-itself that is given by the relations of production. Instead, class as I propose it here is not a set of commonalities, or an underlying unity, but rather the fabrication of a different identity—an identity in struggle—to which individuals and movements can be articulated as part of a political project. (Similarly, it is worthwhile to maintain the centrality of the notion of "capital," since it foregrounds the notions of economy, accumulation, exploitation, and reification. However, it is important to expand this concept to include the domains of power and sovereignty in order to show that the reproduction of social life is not only a matter of the passive activity of a system but also the effect of the assertive application of force and enforcement of order.) Freire's fluid sense of the oppressed suggests a useful framework, though it needs to be developed into an explicitly global project that expresses a positive content and identification.

If the language of class can be useful as a way to understand and organize struggles against colonization, exploitation, sexual violence, and other forms of oppression, it will have to name something other than a pregiven historical dialectic that is working itself out objectively in relation to transformations in the mode of production. At the same time, however, it is important for an oppositional class to preserve the essential sense of struggle—i.e., to be born from the negation of suffering and domination. An effective class subject must be partly defined through its negation of oppression while also going beyond this negation and proposing itself as something more. This complex movement is captured by Frantz Fanon's proposal at the end of *Wretched of the Earth* for a new humanism, both opposed to and exceeding the miserable humanism of Europe, which "for centuries [has] stifled almost the whole of humanity" (1963, 311). Fanon's proposal is paradoxical, since the language in which it is framed is initially given by the European traditions it rejects. But a stronger, newer, and more sensitive sense of class struggle must be prepared to inhabit exactly this kind of paradox, since only this complexity is true to the contradictions of experience and to the actual movement of history.

I suggest that this new class project might be thought of as defined by a *terran* identity.[5] The term "terran" indicates the essential globality of any effective oppositional class, as well as the situatedness of human struggle within specific geographies and more generally upon the earth itself. A terran class might be thought of paradoxically as a particular universal, or a bounded totality, since the whole that it names is both the whole space of our being and history and the very specific and singular globe that we call earth. The emphasis here is not cosmological but geographical, reflecting the essential kindredness of humans and human experience across the shared space and time of life (Shiva 2005). The terran would name a series of material dimensions, from the economic to the anthropological, including the biological, the historical, the cultural, the spiritual, and the political. In addition, this idea foregrounds a fundamental antagonism to the deterritorializations and dispossessions of capital. This conceptualization returns as well to the early Marx's (1964) notion of "species-being," or the creative essence of humanity, which is estranged within the labor process and social universe of capitalism. Capital then is an outrage not only against democracy, justice, and freedom but also against the very materiality of what human bodies grow and strive to do as bodies in life and in history—to create, to imagine, and to know each other authentically.

This oppositional class would bring together the diverse cultural productions and identities of peoples against the monocultural and even *anticultural* homogenizations of capital as consumerism and productivism. The terran might be thought of as a proletarianized version of the currently fashionable "cosmopolitanism" (Appiah 2006; Pollock et al. 2000)—a globally hybrid space characterized by ramifications, migrations, and unsettled identities of travel and exile. The terran class reclaims the materiality of territory against the colonizations of space by sovereignty, and reclaims the process of creation from the logic of capitalist accumulation. It enlarges and extends the humanism of Marx and Freire and situates this humanism as autochthonous historical and cultural production of living bodies. In short, it is one possible name for the new humanism that Fanon proposes but nevertheless leaves nameless—the break and beyond of a history of violence stretching across time and continents.

Praxis and the Terran

Leslie Sklair (2002) argues that the globalized "transnational capitalist class" needs to be defined flexibly and that it comprises not only owners of capital, but also managers, professionals, and bureaucrats who identify with and participate in the project of capitalist expansion and control of the economic, political, and cultural lives of people worldwide. Likewise, our conception of the global class of the oppressed needs to be expanded from the traditional idea of the proletariat to include all those who are exploited, marginalized, or dominated by the power of capitalist globalization. This domination includes not only the incorporation of populations into the global work force but also their expulsion and abandonment—as well as violence done to traditional forms of life, the seizure of resources, and the ravages of war. This definition of the terran is expansive, to say the least, and means that the notion of class has to be imaginatively stretched to encompass a large diversity of social situations and forms of life. At the same time, it remains situated and territorialized in each of its instances rather than being merely a political abstraction. The struggles of the terran class would include, for example, the fight for traditional land use rights by indigenous groups, the battle against cutbacks and downsizing in manufacturing, and movements exposing and challenging the convergence of sexual violence and economic exploitation, among many others. The praxis of this class has to be thought of as a set of tendencies rather than a single program. Rather than focusing on the creation of formal institutional links, this praxis would aim primarily to develop a global consciousness and sense of affiliation as a foundation not only for loosely coordinated actions but also for individual local struggles as well. This suggests a global project of "conscientization," to use Freire's (1997) term, oriented toward the scale of the totality. The first task of this conscientization is a recognition and analysis of power and capital as antagonists.

Theorists have pointed to the importance of new forms of political organization in the global era. Rather than nation-centered labor unions or political parties, the present conjuncture sees, for example, the proliferation of issue-oriented NGOs that are sometimes transnational in their membership (Sklair 2002). More dramatically, Appadurai (2006) describes an emerging "cellular" form of organization that characterizes both reactionary and progressive social forces (e.g., both Al Qaeda and left grassroots movements). He argues that this cellular form of organization, in contrast to the traditional "vertebrate" form (which characterizes the structure of nation-states and their international protocols and agreements), is decentralized, loose, adept at replication, and reliant on high-tech communications. This is an evocative description, and certainly many global protest movements in recent years can be usefully characterized in these terms.

However, the construction of a terran political project, while similarly decentralized, would be a more public and pedagogical project aimed less at the ramification of a particular organization and more at the elaboration of a common and critical vision of globality. On the other hand, while dialogical, this project would be more than a mere opening to a process of deliberative democracy.[6] The terran proposes a common, if loose, form of global identification, as well as a political force opposed to the destruction of authentic human and global relationships—a force ready to act against neoliberal and imperialist projects worldwide and against the innumerable instances of state and corporate power directed toward the violent erasure or assimilation of resistance.

Such a praxis will begin to be realized when local struggles start to see themselves in others elsewhere—just as, in their own way, disparate forms of domination are in fact deeply linked. For example, the "accountability" movement in education mimics, on its own terrain, the structural adjustment that neoliberalism has forced on developing economies through the agency of international financial institutions. Pauline Lipman (2004) shows how patterns of school reform and gentrification in Chicago, designed to restructure urban space in favor of white middle-class elites, are connected to stratification and polarization processes common to global cities generally. The patterns in Chicago are typical of many urban spaces and school systems in the United States, and yet it is still rare that the political discussion around this restructuring is framed in terms of globalization and transnational capital. Interestingly, another contemporary political struggle initially centered in the sphere of education—the 2006 strike of teachers in Oaxaca—grew quickly into a popular uprising comprising many sectors and organizations of civil society, who organized themselves into the Popular Assembly of the Peoples of Oaxaca. But beyond even this regional articulation, what would it look like to take the struggles in all of these locations one step further, and through the recognition of common goals and opponents to create a global identification in resistance? What would it mean for community members, students, and teachers in Chicago to take the bearings for their projects not only from the immediate assaults they confront but also from the context of a transnational project of opposition? If this is to be more than a mere trading of information or ideas, then a common class project and identification must be built, even if it remains decentralized.

The notion of the terran recognizes both the situatedness of individual struggles and their participation in the larger ecosystem of the global. On the one hand, like Hardt and Negri's multitude, the terran foregrounds the transnational dimension; on the other hand, like Freire's conceptualization of the struggle of the oppressed for humanization, the terran emphasizes the importance of human relationships and their determined opposition to domination. In this conception, solidarity begins to weave together a self-conscious historical subject that is more than a simple coalition, although it is less than a unity. While the praxis and goals of this subject are political and social (as in Hardt and Negri), they also have important ethical and spiritual dimensions (as in Freire). This is because authentic revolution, and the authentic construction of a new mode of social life, are not only about imagining a new *polis* but also about constructing a new culture and human community. This community would be grounded in a respect for and attention to the earth (where this refers not only to "nature" but to human beings as inhabitants of the earth—beings embodied and ensouled, one might almost say *telluric*) while also being directed toward the construction of a new social world. Dewey's (1944) conception of democracy as involving the proliferation of relationships, identification with civic space, and emphasis on productive collaboration would be projected globally in this conception. But as a revolutionary project, this democracy would be built in the course of a faithfulness to the critical imperative of struggle against subjugation and exploitation.

Conclusion

The flexibility of power in the present and the diversity of struggles against it call for a creative thinking through of the possible convergences of global oppositional movements, as contemporary observers such as Harvey and Mohanty suggest. The critical and Marxist traditions offer powerful resources in this effort, which must be mined eclectically in order to recognize the useful in the new without prematurely dismissing what is of value in the familiar. In this way, Hardt and Negri's evocative description of the multitude can become an important starting point for a new political project to the extent that this project is reterritorialized, grounded in the experience of struggle, and made sensitive to the ongoing power of the negative in the context of the persistent marginalization, violation, and subjugation of people throughout the world. Freire's dialectic of oppression and liberation, from which emerges the central contradiction between oppressor and oppressed in contemporary societies, continues to be indispensable as a framework for making sense of power in the present and resistance to it. Nevertheless, Freire's model needs to be rearticulated to respond to the scale of the global and must also take the step of naming a positive project within and against ongoing social violence. To respond to these imperatives, I have proposed here the idea of a *terran* class, which would bring together the full range of those who are damaged by capital within a fluid project of solidarity and identification. This class identification would not supersede local determinations but would rather name the total set of relationships that organize both local struggles and the panorama of the global. This class could begin to challenge contemporary forms of rule and exploitation that are already fully global in their logic and extent.

The evaluation of gains and losses in this project has to involve something more than recording successful defenses against particular assaults. It also has to be measured in the development of networks of communication, mechanisms of coordination, and new subjective formations. Whatever individual defeats take place, if a globally integrated culture of resistance can be built, then these individual losses will be better sustained, and a horizon of possibility will be available that will serve as an indispensable resource for movements everywhere. For young people growing up in a very uncertain world, the development not merely of a global consciousness but of an oppositional global identification is urgent, as is the collective action, on a planetary scale, that can alone afford the resource of hope they will need to confront a very difficult historical moment. In this regard, the pedagogical dimension of this larger philosophy of praxis is central. The challenge for critical education is to make vivid for students their own actually existing relationships, in their various forms, to global processes of power and exploitation and on this basis to reconfigure identifications in terms of solidarity and kindredness with those in struggle elsewhere. After all, if the imaginative visions of the theorists

I have discussed here, and my own proposal for a global oppositional project that I have described above are to become real in any sense, they will have to be taken up as their own by those who will inherit both a deepening crisis within capitalist society and perhaps the best opportunity yet for fully confronting and overcoming it.

CHAPTER 7

Globality, Globalization, and Critical Pedagogy

For critical pedagogy, confronting the terrain of the global means considering not only implications of the objective reorganization of social life as a result of new transnational economic and political regimes but the implications also of a dramatic transformation in the context for critical teaching and activism that the scale of the global, as a fundamental frame of reference, represents. While the previous chapter considers how oppositional movement needs to be reconceptualized in the present, in this chapter I am concerned with the impact of globalizing processes on identities and relationships, and particularly relationships within educational contexts. These processes alter the underlying settings for interventions, and critical pedagogy must rethink its strategies within these new settings. If the nature of oppression is different, or if its prevalent modes or experiences are altered, antioppressive education must be prepared to adapt as well. And if the meanings and possibilities of "democracy" are deeply affected by globalization, forms of pedagogy oriented toward creating democratic spaces and relationships must be attentive to these effects. In this chapter, I discuss several of the most important aspects of globality and globalization as they relate to the reorganization of social and political life and the possibilities for democratic struggle, giving special attention to changes that they suggest for critical approaches to teaching and for the identities of critical educators and activists themselves.[1]

Common usage of the term *globalization* conflates two different meanings. On the one hand, this term is shorthand for a specific set of economic and political initiatives undertaken by global elites as part of a new phase in the history of capitalism. In this usage, the term refers first of all to neoliberalism—the global disciplining of workers, the poor, and developing societies in order to respond to a crisis of accumulation in the leading capitalist societies—though secondarily it refers as well to the spread of transnational corporations and consumerism more generally. On the other hand, globalization is also used to refer to a more fundamental process of the withering of the nation-state system as the primary framework for organizing social and political life and the worldwide cultural interpenetration that reorganizes human society and identities on a planetary scale. This second sense of the term is existential as well as political, and it is more or less synonymous with *globality*. Of course, we live in a particular world, one dominated economically, politically, and culturally by very particular elites, which means that even if these two senses of globalization are different on paper, by and large they overlap in fact, since our experience of the shift to globality is essentially mediated by the powerful and by *their* vision of what a global society can and should look like. And, of course, to the extent that the development of capitalism is the engine that has historically driven the reconfiguration of human life on a vaster and vaster scale, it is not surprising that globalization should be experienced as a new set of experiences of production and consumption.

Both of these meanings and dimensions of globalization pose dramatic challenges for people everywhere. First of all, globalization has so far meant, among other things, the decline of stable jobs and good benefits for many workers, the proliferation of conditions of super-exploitation for others, the abandonment of many to no livelihood at all with the dramatic movement of firms around the globe, the destruction of traditional economies and forms of life, forced migration, cultural imperialism, and predatory consumerism, not to mention environmental degradation and perpetual war. For children and youth, these processes have been devastating, leading to important threats to survival and stability. For young people in the global North, these are associated with the specific challenges of incarceration, violence, and social marginalization; in the rest of the world, destabilizations related to globalization have made children the targets of war, child labor and slavery, conscription into child armies, and new pandemics. But beside these problems, which are associated with the first, directly political-economic, sense of globalization mentioned above, there are additional, if less tangible, challenges associated with the second sense of globalization—that is, the shift to the condition of globality itself, the organization of human life and meaning on a much vaster and more complex scale. At the most basic level, this movement toward a global organization of social life means the interruption of local narratives and expectations, as well as the experience of powerlessness in the face of apparently vast historical forces. It is also associated with the replacement of familiar frameworks and modes of communication by alien ones, the deterritorialization of identities, and the assimilation of daily practices to a new set of general and planetary social habits. If these are dramatic changes for people generally, for young people they are particularly stark, unmooring them from familiar contexts, teaching them extreme forms of alienation, and throwing them headlong into the coldness of a future with no guarantees.

Of course, many observers have pointed out that globalization and the condition of globality also create new and important social possibilities and opportunities. The very insatiability of power and the incessant expansiveness of capital, as they remake the conditions in which people work and live, driving them ever more completely into the culture of the commodity, also create a new kind of commonality between people everywhere. This means that while people increasingly share in the experience of subjugation to the same free-market fundamentalism, they also potentially share in new forms of oppositional identity. In addition, the spread of popular culture around the world (even if not in all directions equally) potentially makes possible new and powerfully hybrid forms of art, politics, and identity. The tools of the powerful, in particular telecommunications and the Internet, are to some extent also available for global social justice movements. And it may be that Marx and Engels (1848/1967) are still correct in their view that there are certain forms of parochialism that it is a blessing rather than a curse to be made free of through the influence of the spread of capitalist culture and the struggle against it. It is important, at any rate, to point out that if there are positive possibilities that emerge through globalizing processes, they are important precisely in being creatively discovered *in resistance* to actually existing globalization. This accounts for the complex identity of the new transnational protest movements, which are both antiglobalization and pro–alternative globalization at the same time.

If young people are especially exposed to the dangers and challenges of globalization described above, they are also at the cutting edge of the opportunities it presents. After all, if these opportunities are to be taken advantage of productively, and if the threats to sociality and survival that are posed by global immiseration, war, and plunder are to be countered, it is young people who will do it. Thinking carefully about education and pedagogy is crucial in this regard. Furthermore, given the reorganizations of experience and identity produced by globalization mentioned above, it is clear that teaching and learning as part of the movement for social transformation must be reconceptualized in some fundamental ways. Teaching for social justice must amount to something more than recognizing the new evils of the rulers. In addition, a critical pedagogy of globality and globalization must be able to reckon with the fundamental transformations of consciousness, experience, and identity that are central to the shift to the historical condition of globality. Not only does this mean a consciously transnational perspective, but it also implies a flexibility and innovativeness that can respond to the terrifying (and potentially exhilarating) openings that the landscape of the global forces upon us.

Globalization and Identity

Changes in the objective structures and conditions of social life are deeply intertwined with changes in the kinds of meanings that can be constructed to give sense to the lives of individuals and communities. Globalization puts familiar forms of identity under pressure, as people are variously marginalized and incorporated by new economic and political processes. Widespread immigration, itself an effect and crucial dimension of the global economy, challenges given identifications as well as provoking new ones. For example, Saskia Sassen (1998) describes how immigrants in the United States are incorporated into a new "serving class" upon which depend the elite beneficiaries of the global economy. At the same time, the feminization of service work gives women in this sector access to income and independence that they often did not have before. Globally, gender identities and worker identities come together in ways that are empowering (as when women make use of gendered modes of communication and organization in the service of workplace solidarity) as well as disempowering (as when capital takes advantage of the devaluation of women's work in order to increase the rate of exploitation)—see Mies (1998). Beyond this, globalization puts in doubt the validity of national identifications generally, as the national space is penetrated by supranational economic and political processes that throw disparate populations together in new ways while revealing in especially dramatic fashion the fundamentally imaginary nature of national identity in the first place (Anderson 1991). Thus, the diversification of the population in the United States challenges white-supremacist assumptions about the cultural content of "Americanness," which is expressed in new forms of linguicism, xenophobia, and racism. In this connection, Arjun Appadurai (2006) describes the difficult dialectics of "majority" and "minority" populations within the context of the insecurity provoked by globalization. Ethnic cleansing and genocide are catastrophic attempts to respond to this uncertainty, as "majority" populations seek to exorcise the implicit threat posed by minorities and demographic diversity generally—namely, the possibility that these majorities represent merely a contingent demographic reality rather than a pure expression of the nation.

Globalizing processes reorganize the basic conditions for being and understanding the limits of oneself and one's context. Globalization literally and figuratively deracinates and deterritorializes people, throwing them out of occupations that gave meaning to their lives and disrupting communities and cultures. But it also powerfully articulates individual lives to global forces and through them to the terrain of the global itself. The North American Free Trade Agreement has wrought havoc on indigenous communities in Mexico, while also provoking them into forms of resistance that directly confront not merely the local *cacique* but the heart of globalizing capital itself. Furthermore, this resistance serves as a crucial node in a worldwide set of popular movements of opposition. In this way, alongside the prevailing mode of globalization as the increasingly total subjugation of the social to capital, there is as well a dramatic alternative and oppositional vista on an absolutely different scale from the old visions—the scale of the totality. In addition to the traditional vertical projections of identity upwards from the soil of community, ethnos, and nation, there is the possibility of a horizontal extension, as people are linked sectorally to others resisting the same social forces and conditions across the globe. This is a dramatically new reconfiguration of class identification and struggle as the vast majority begin to find a new solidarity against an increasingly embattled global elite. The movements of the dispossessed in South Africa, as Ashwin Desai (2002) reports, have given rise to an original identity, outside of and opposed to the party, ethnic, racial, and workerist ones of prior struggles: the "poors" are all those who find themselves sharing the common condition of subjugation to a neoliberalizing economy managed by a "revolutionary" capitalist party (the African National Congress). But this is not simply a negative identification, the designation for those who have been left behind by neoliberalism; this is also potentially the banner of a powerful new social subject. The poors stake a claim to the empty space laid bare by globalization—the space of the discarded, redundant, and marginalized. History, they say, belongs as much to them as to the rulers and managers.

Of course, there are also less oppositional articulations of the opportunities opened up by globalization for new identities, subjects, and citizens. The boosters of capitalist globalization tout in particular the exciting forms of empowerment available to the emerging middle classes in developing countries. Thus, Thomas Friedman (2005) is ecstatic about the upward mobility of Indian service workers employed in call centers and high-tech firms. He is silent with respect to the violence of capital, except to the extent that he notes that some populations have not yet figured out how to connect to the benefits of the new "flat world" of globalization. A more sophisticated version of this optimism is promoted in the currently fashionable idea of "cosmopolitanism," which argues that globalization makes possible a kind of transnational citizenship that values cultural difference while also promoting certain ethical universalisms (Appiah 2006). This proposal glibly runs together the vast range of different confrontations with the global; the intellectual's leisurely appreciation for the varieties of human experience is different from the peasant's sudden apprehension of a globalized economy that forces him or her out of a livelihood. But the idea of cosmopolitanism does foreground an important new form of agency, if properly appropriated. The power of protest on the global scale is made possible, after all, by a truly transnational and cosmopolitan effort of communication and coordination, even if the effects of this protest remain so far uneven and difficult to assess.

Marx and Engels (1848/1967) argued that the conquest of the world by the capitalist mode of production was a brutal and tragic passage, as well as an enlightening and productive one since it opened a broader vista for the vast majority and made possible a general human and historical progress. This perspective may too quickly dismiss what has been lost and damaged. It is wrong to assume *a priori* that the historical dialectic is ultimately worth the ravages of its unfolding. By the same token, however, it is important not to overlook the powerful possibilities made available by globalization as a general process of cultural transformation. It is important to explore the new windows that are opened on and for human being by this basic shift in the conditions of existence. The global perhaps unfolds a new technology of the human, of which the glossy new software and communications technologies are only a weak reflection. This deeper formation would not be an instrumental one but rather a technology of *solidarity*—a more powerful and liberatory organization of human relationships across the planet, capable perhaps of finally contesting the subjection of sociality to the imperative of accumulation (De Lissovoy and McLaren 2005). This solidarity is as yet only emergent. It will require a great deal of physical, mental, and spiritual work to accomplish it. And it will require a deeply pedagogical engagement, since it cannot simply be manufactured; it can only be collaboratively learned and communicated. This is the central historical task, especially for the young. For educators, this means participating in a process of working with students through successive challenges and anxieties, as familiar frames of references are replaced, new relationships formed, and new knowledge gained. And perhaps even more profoundly than is usually imagined (even by critical educators), this will have to be a process in which teachers are learners just as much as students are. Rather than the expert teaching the novice or the leader guiding the disciple, the global itself, as overarching condition and horizon, teaches everyone together a new form of life.

The "Personality" of Globalization

In capitalism, according to Marx, the labor process "posits the real objective conditions of living labor . . . as alien, independent existences—or as the mode of existence of an *alien person*, as self-sufficient values for-themselves, and hence as values which form wealth alien to an isolated and subjective labor capacity" (1973, 461). In the present, it is important to recognize this process not only within the narrow confines of economic production but also throughout global society. The totality of social life, reproduced on a grander and grander scale, increasingly appears to be alien to the individual, to be directed from "elsewhere." This is a basic characteristic of globalization as well as the new imperialism. As capitalism responds to its internal crises through expansion and extension, this results in a vast reproduction of the stuff and circuits of the economy and also in an accumulation of the depth and *distance* of alienation itself. Power and capital in the present should not be described exactly as tyrannical. Rather, they are mediated by a profound indifference, which removes the origin of social violence from those who experience it, quantitatively across space and society as well as qualitatively, since the purposes that organize this violence are those of an ever more remote subject, ever more arbitrarily applied. In this way, transnational firms seek to control the supply of seed traditionally used by farmers in India (Shiva 2005), and the executive branch of the United States government invents novel interpretations of the law to permit it to designate people of color from Chicago or Kabul as "enemy combatants" at will and to deprive them of all due process. The same lack of authentic accountability is ultimately at work in the official "accountability" initiatives in education, as every child in the United States is subjected to standardized norm-referenced tests that register his or her existence according to only the most instrumental and arbitrary problematic. One of the objects of resistance then becomes the defeat of this distance and an assault on the barricades that protect the center, as in the dragging down of the police fence by protesters at the meeting of the World Trade Organization in Cancún in 2003. Such protests are threatening in terms of the interference they create, but they are perhaps even more so in the literal *approach* to power that they represent—the traversal of the geographical, political, and existential space that separates ruler and ruled.

In the context of the international division of labor, it is the continued exploitation of workers in the global South that makes possible the glut of capacity and goods in the North, however complicated the networks of the postmodern economy of late capitalism. But it is important to recognize that beyond these commodities themselves, the total reification of social life and the dizzying panorama of contemporary consumer culture are themselves the product of the alienation of the creative energy of the mass of humanity; in a deep sense, this spectacular social production belongs to the mass. Even the economic, political, and ideological resources that construct the power of elites are themselves derived from a monopoly on the creative force of human labor power and subjectivity, a force that belongs ultimately to the people. In other words, exploitation does not merely pay for the continuation and expansion of capitalism and imperialism. Exploitation also alienates the collective human will, energy, and "species-being," and seeks to reserve for the rulers not only the vocation but also the resources, in an existential sense, to make history (Freire 1997). Not only are the jets and bombs, the limousines and grand habitations of the powerful, as well as the space and air for them to pontificate and excoriate, paid for in the sweat and blood of regular people; in addition, the very offices, vocabularies, and identities of the powerful—the very possibility of the selves they activate and mobilize—are *made out of* the lives of the oppressed. They are the material that the consumption of these lives produces.

In globalization, just as in the classical period of capitalism, the objective conditions of labor, in their alienation from the subjectivity of the worker, acquire an alien subjectivity and take on what Marx called a "personality."[2] In the same way that the concept of capital necessarily includes the capitalist, globalization itself, which represents the almost absolute reach of commodity production and militarism on a world scale, also necessarily produces its own personality. Globalization is not merely the working out of an economic or political problem by objective and abstract forces. It is also the emergence of a historical subject that we must be able to recognize. The anger of people at violation and exploitation is not directed at abstract principles of economics; it is focused instead on a dynamic constellation produced in the popular imagination, which is embodied in the persons of the powerful but also operates beyond them. We need to delineate that historical subject in order to give shape to what is already felt and acted upon in the popular consciousness. In the United States, the prominence of the current dominant political fraction (i.e., the neoconservatives) should not blind us to the ways in which this group expresses a larger consensus among the powerful, and yet the pronouncements of this fraction do end up becoming an idiom for imperialism itself. This dialectic is creatively represented at antiwar protests in the huge masks and cardboard cut-outs of the figures of George W. Bush and Dick Cheney. These masks allow us to put a face to globalization and the new imperialism; that is, to approximate its actual class-subjectivity with the images of particular living subjects.

Focusing on the "personalities" of globalization may partly conceal the systematic and impersonal nature of exploitation. But images also work to provide a coherent target that can constitute an opponent for movements. And popular understandings of this opponent are crucial contributions to the genuine understanding of the present. We should look for ways to develop our representations of globalization and liberatory movement as countersubjects through a tracing of the existential register of their contest. Students, teachers, and activists should consider the firsthand testimony of those in Iraq and elsewhere who are suffering the worst assaults of current military adventures, while also considering the connections between different dimensions and moments of oppositional struggle—both to deepen the understanding of the whole and to bring different constituents together toward a more powerful movement. Our knowledge should proceed from the informed intuitions and impulses of this movement itself. Theory itself is only effective to the extent that it has acquired what Gramsci (1971) called a "practical efficacity" and learns to contribute materially to the actual process of struggle.

Global Power and the Possibility of Democracy

Globalization provokes questions about new dimensions of power as well as the challenges and possibilities for democracy; critical education must analyze and explore these questions. Is the essence of power in this historical moment the same as before, or different? Can democratic principles that already exist be simply extended to the scale of the global, or is it necessary to invent new ones? In responding to these questions, an important conceptual starting point is the idea that capitalism, and thus society generally, is facing a deep and global crisis of overproduction—an inability to find enough profitable outlets for the surplus capacity and capital that has been accumulated in the process of production (Brenner 2002). This is the source of the never-ending search for cheaper and cheaper labor and is the motivation behind neoliberalism and its impulses to privatization. To be able to continue to expand, capitalism must penetrate spheres that have so far been external to it. First of all, this involves expanded reproduction (the enrollment of more and more people into the productive process as workers). But in the context of systemic crisis, more drastic means must be found—including simple plunder, in which public resources and economic sectors are commandeered by transnational firms (for example the privatization of utilities, water services, and transportation) and human creativity and the riches of nature are commodified in new ways (for example the patenting of natural organisms and indigenous knowledge). At work here, then, are the simultaneous and contradictory processes of proletarianization and deindustrialization, as some are incorporated into punishing factory work while others are expelled from it and exploited in new ways, and as capital searches for previously uncolonized areas of social life to penetrate (Harvey 2003).

The dramatic ramification and complexification of the global economy is a fundamental fact that affects everyone and that must reorient critical efforts to understand and intervene against power. For one thing, the very notions of democracy, the public sphere, and hegemony, which have served as basic organizing principles for critical educators and activists, have to be reexamined in light of global processes that do not necessarily respect the political limits and logic that organize these ideas. Globalization tends to absorb all public spaces and processes into the logic of capital without regard to national boundaries, variously extending or withdrawing productive capacity at the same time that it commodifies culture everywhere. Therefore, globalization undermines the usefulness of political strategies organized around, or conceived in the context of, the nation-state. In broad terms, as Samir Amin (1997) describes, this is the political crisis that confronts contemporary societies, as ideologies and languages across the spectrum that are concerned with paradigms and policies of the nation-state fail to come to terms with a global economic (and social) reality that is not premised on this prior political logic. To the extent that critical pedagogy, as well as popular movements, conceptualize their projects in terms of discrete national spaces and in terms of building historical blocs capable of intervening to influence the state, they fail to adequately comprehend the present and risk being sidelined by contemporary developments.

By the same token, understanding globalization can shed new light on many of the most intractable issues that we confront. For example, the effort to reduce and privatize social services in the United States (including education), which is usually criticized simply as the ill-conceived project of ideological conservatives, can instead be understood more broadly as a form of structural adjustment designed both to reorient public life to the culture of the market and to absorb new sectors of it into the sphere of capitalist accumulation (McLaren 2000; Saltman 2005). However, we should go even further and recognize that we have to do here with processes that are more than economic in the narrow sense. The new colonizations and penetrations of the era of globalization are also *biopolitical*—that is, they aim to assimilate and exploit not only labor power, as traditionally understood, but even the fundamental capacities of desire, communication, and affect that organize human sociality in the first place (Hardt and Negri 2004). Thus, the effect of recent transformations in education (for example, the spread of voucher and "accountability" initiatives) is not just to inculcate ideologies but also to *reorganize* the very subjectivities, habits, and desires that construct who and what students might be as social beings. And globalizing processes are also, I would argue, forms of spiritual plunder, as the hope and solidarity of humanity are alienated into a profound despair that seems not only impotent against power but also reproduces forms of social crisis that appear to justify power's intervention and management of social life.[3] This dynamic is at work first of all in the expulsion of young people from the worlds of education and work into conditions of hopelessness, and then again, in response to this problem, in the invention of "solutions" in the form of new non-spaces for youth to inhabit as mercenaries on the proliferating fronts of the global war or as inmates in the forgotten landscapes of the prison-industrial complex.

What notion of democracy can be adequate to these difficult conditions and to the very framework of the global? Some have argued that the historical goals of socialism—the overcoming of capitalism and the establishment of alternative forms of international social production—must remain the goals for any democratic movement

(Amin 1997); in fact, capitalist globalization makes this a realistic and urgent project in a new way, as transnational forces and projects of opposition can be more clearly observed and imagined. On the other hand, others have argued that while the global era involves an increasing interconnectedness of peoples and recognition of legitimate differences, the basic principles of deliberative democracy remain unchanged. In this view, the consensual agreement of equals, as Seyla Benhabib (2002) puts it, arrived at within conditions of universal respect and reciprocity, is the precondition for truly democratic politics, and perhaps especially so in the context of a global multiculturalism. Others have argued that an entirely new paradigm is needed and that democracy can only be imagined as the political project of a new global actor that we can only begin to glimpse in the present. Thus, for Hardt and Negri (2004), democracy is simply the progressive materialization of this subject—a deepening and widening of the networks of social communication and collaboration of the "multitude." However, any understanding of democracy in the age of globalization, and any critical pedagogical project to imagine and construct it, must come to terms with the fact that discourses and strategies based on narratives of the nation and logics of the national state must give way to a global conception. This means recognizing that globalization and globality represent more than an extended internationalism and instead constitute a fundamentally new mode and horizon of social life.

New Collectivities, New Struggles

The patterns and changes described above together create a condition of extreme uncertainty. The precariousness of life chances, the erosion of norms and expectations, the pressures on familiar forms of identity, and the distance and velocity of power confront individuals with daunting challenges. Zygmunt Bauman (2000) calls this condition "liquid modernity," a historical stage characterized by a new fluidity and impermanence, as opposed to the solidity of the structures of state, society, and rationality that characterized the old order (or "solid" modernity). Furthermore, as Bauman describes, at the same time that individuals are faced with fundamentally new uncertainties, they are also given the sole responsibility for navigating this landscape and weaving together some reliable framework of meaning and security—even though the difficult conditions they face are in fact the result of systemic contradictions. Therefore, the task of critical theory and pedagogy, and radical politics generally, is not only to expose and resist the organized power of the state and society but also to begin to imagine a different sociality. The problem is not merely the way that power intrudes into the lives of people but in addition the way that it retreats and abandons individuals to their fates. Neoliberalism perfects the catastrophic synthesis of these two projects, assimilating populations into the process of capitalist reproduction on an expanded scale and then just as suddenly expelling them into the gigantic global reserve labor army as firms scale down or move across the globe. Even in societies at the center of global power, the growth of a new, many-tentacled security state coincides with a tendency away from the forms of control associated with the extended welfare state, as the drive for flexibility on the part of capital results in the paring down of state regulations and interventions in the economy. In an older sociological idiom, this is both a decolonization and recolonization of the "life-world."

In this context, while it is necessary to reinvigorate the public sphere and promote forms of enlivened and engaged citizenship and pedagogy, as Bauman (2000) and Henry Giroux (1988, 2003) argue, this is not enough. If global power aggressively intervenes in social life at the same time that it flees society (and its commitments to it), then this power must be assertively contested and not merely bypassed. This means that principled critique and confrontation remain crucial responsibilities for the left in theory and practice. In addition, calls to revivify arenas of public life and democracy, including education, must be careful not to fall back on the very senses of subjectivity that power relies on as mystifications. A project of global social transformation means more than simply supplying actual content to the empty ideologeme of the "individual." Against the idea that the critical task remains "the self-constitution of individual life and the weaving as well as the servicing of the networks of bonds with other self-constituting individuals" (Bauman 2000, 49), we must consider whether globality at last presents us with the possibility of escaping individuality as the organizing logic of social life and with the chance to discover a new logic of collectivity. This is an important challenge for critical pedagogy to the extent that it has tended to emphasize conscientization at the individual level against the deterministic and monolithic senses of agency of the old left (e.g., Freire 1997, 1999). But if familiar ideas of class and collectivity belong to an older modernity that is fast being eroded, what new classes might be imagined or built, or might perhaps already be emerging? If "democracy" can be rethought in ways that are able to substantively respond to the dilemmas posed by globality, it must be rethought in this context.

A dramatically fluid and volatile present, which remakes the conditions of social life, meaning, and identification in an instant, calls for an equally radical project of collective imagination and transformation. A new collective subject of opposition and alternative globality needs to set out into the unknown and to discover itself in the process of building another world. That this is indeed a process of opposition as well as creation means that the emphasis in critical theory on public life has to be radicalized, and social transformation conceived of as a project aimed specifically *against* power and capital at the same time as it unfolds a culture of democracy. Social justice movements cannot expect neoliberalism and empire to simply fade away as new transnational collaborations are organized; they will have to be countered and challenged at the same that a different future is built. This contradiction is evident in the gatherings of the World Social Forum, which seem to suggest in their transnationalism a new democratic movement but which at the same time more immediately bring together specific regional struggles against neoliberal assaults (Mertes 2003). Class antagonisms, reconfigured and projected on a global scale, will remain. On the other hand, even in the context of this struggle, the fundamentally new conditions of the present will require a radically creative imagination, one that goes beyond, for instance, a Deweyan and radical democratic perspective. Rather than a project of social *re*construction (Dewey 1944), which implies a redoing (and doing better) of something that has already been accomplished, what is necessary in the present is the building of a form of life that is original. This is an open-ended project, since not only can the future not be known before its construction, but in addition the subject that creates it can only gradually be disclosed to itself in the process.

What new class and class consciousness will emerge from the ranks of the marginalized, the dispossessed, and the increasingly vast majority that suffers the depredations of the global elite and the social processes that reproduce its hegemony? The shackdwellers' movements, the organizations of part-time and unemployed workers, and the unions of the landless and the cast-off do not easily fit the old class categories, and yet they represent powerful new forms of political agency and subjectivity; these extend the challenge by the social movements of the latter part of the twentieth century to the hegemony of labor within a left opposition (Aronowitz 2006). The Global Women's Strike highlights the invisibility of the labor of women and girls as well as the global inequities in pay and life chances between women and men, and in so doing strikes at the heart of capital, which depends on this invisibility both to hide and to ratchet up the rate of exploitation. A widespread and widening ecological consciousness, to the extent that it is serious in its commitments, must come up against the imperative of capital to constantly expand (whether in traditional or "green" industries, as Joel Kovel [2002] points out) and therefore must begin to create strategies against capitalism itself. These creative experiments, and others, begin to expose new political landscapes and to suggest new forms of subjectivity and sociality, ones not based on the exchange value of human creativity or on the exploitability of natural diversity. Critical pedagogy must urgently make these emergent tendencies available to youth. If these new movements are still somewhat open and amorphous, that is because they are outside the social logic of capital and so can only appear as blank spaces in the global geography that capital has incessantly mapped and mined to exhaustion. The actual shape they come to take, and the names of the subjects they propose, can only be spoken in the actual unfolding of their organization and opposition, a process in which young people must necessarily take a central part.

Teaching and Solidarity

If social transformation at this point in history can only be imagined in global terms, as the foregoing discussion suggests, this means that liberatory education has to have a perspective on the relationship of local and global, and a strategy for mediating this dialectic. Critical pedagogy in the present should make vivid for students their own actually existing relationships, as inhabitants of a territory or region, to broader relations of power and exploitation, on the model of the concentric regional circles of Paulo Freire's "generative themes" (1997). On the basis of this interpolation of the global into the imaginary and insular space of the national, students can begin to reconfigure identifications in terms of a solidarity with those in struggle across the globe. But in order for this to take place, it is also necessary for educators to engage students in a critical reading not merely of injustice or oppression generally but of power and capital as global processes. This does not mean imposing a distant and alienating vocabulary but rather initiating students into a new mode of thinking about their own lives and communities. At the same time, however, this may in part feel like the intrusion of a dangerous discourse, especially in the United States. But in this pedagogical and discursive choice a political one is made as well: to participate in a conversation that has been joined by radical and left movements globally rather than to surrender to limits tacitly enforced by a parochial progressivism.

But at the same time that critical teaching connects students to urgent global questions and to a critical reading of power, it must also rethink its own assumptions. The paradigm shift involved in the transformation of national and international issues and identifications to global ones means that critical pedagogy must also be transformed. In particular, in a social universe whose basic realities are changing at an unprecedented pace and in which there is essentially no simple map by which to navigate, the function of the teacher as leader or guide is challenged. For better or worse, in a world crisscrossed by unprecedented and expanding fissures in its social, political, economic, and ecological fabric, young people themselves will ultimately be the ones who discover a path. This does not imply that educators should *abdicate* their place and power; rather, it means that critical teachers must discover ways to initiate and collaborate in this process with young people rather than imagining themselves as in possession of a fundamentally different and deeper understanding and authority. Human beings are faced now not simply with the responsibility to pursue the vocation of humanization and to overcome oppression; beyond this, they confront the necessity of constructing a new world. The imagination of young people must take a new and more central role in this project, since their visions are less likely to be constrained by the limitations of discourses and traditions that were forged in conditions that are disappearing.

Furthermore, in the double-sided movement of neoliberalism described above, new forms of hyperauthoritarianism are produced at the same time that power finds new ways to secede from public space. This is particularly apparent in schools, where new regimes of policy and pedagogy reconstruct education as a form of punishment (Lyons and Drew 2006). This is an entirely different degree of authoritarianism than that with which critical pedagogy has so far concerned itself. Rather than focusing simply on the construction of docile and compliant subjects, this new authoritarianism—as expressed in exit exams, zero tolerance, "back to basics" and scripted curricula, and other measures—aims covertly to exclude vast numbers of youth from participation in public life altogether. In this environment, any rigid consolidation of authority (even "critical") risks being experienced as violent—as well as being assimilated by the hyperauthority that is omnipresent and overdetermining within the institutions. This does not mean that the very idea of authority should be dispensed with. Wherever there is solidarity, there is some form of authority. The challenge is to imagine more fully collective and collaborative organizations of authority, which distribute it across a network of participants. The uniqueness of the teacher's position in this arrangement would not be that of a leader, or of some disinterested facilitator, but rather that of provocateur, senior participant, chief organizer, and mentor. The existential, political, and ethical imperative to learn to navigate a fluid world together means inventing more profoundly collective solidarities than we have known before. In addition to interrogating received wisdoms and identifications in order to produce hybrid classroom cultures, this means participating in the construction of larger and properly global communities and political projects, which can then be enacted in a multiplicity of individual sites, including educational ones.

Rather than initiating students into already established "communities of practice" or socializing students into the established habits of a stable democratic society, the center of gravity must shift to constructing new communities of practice, new habits, and a new society. There is an important shift that must take place from incorporation to creation, from inculcating democratic culture to clearing the ground for the discovery of a new culture for a new world. For example, although it is necessary to recognize, in the context of changing technologies, the importance of cyberliteracies and "multiliteracies" in addition to traditional print literacy (New London Group 1996), educators must also consider what new communities are made possible for young people by the changing landscape of electronic communication and what new forms of abandonment are being invented for those who are excluded from these landscapes. Since the possibilities of the new are discovered in the process of determined opposition to the assaults of power and capital in the present, this also means that as power becomes more aggressive and mobile, so must educational opposition—it must look beyond the classroom and participate in struggles on the ground against global militarism, racism, and privatization. Youth themselves have recently led the way in walkouts and protests in support of immigrants' rights, against the war in Iraq, and against the impoverishment of curriculum and opportunities in the schools. These struggles are at the leading edge not merely of social movements but also of critical pedagogy. They suggest new collectivities, commitments, and kinds of mobility that critical teachers should study and take their cues from as they seek to bring their own resources and understanding into active collaboration with the energy of students, which is already in motion.

Pedagogy and Liberatory Imagination

We can only uncomfortably imagine what is outside the actually existing universe of things from a perspective that is located inside that universe. This form of imagination is a painful kind of thinking. It doesn't have a genre or an institutional sanction. It only truly becomes possible when existing modes of being become impossible. Otherwise, why venture into an uncharted and unrecognizable terrain? It means being more creative and energetic but also risking the structures and identities that make existence intelligible, even if that existence is, at its core, alienated and wretched. It means bursting the integument that allows for an intelligible self at the same time that it refuses the possibility of transformation.

The saturation of life by capital in the present ends up exteriorizing us to the extent that we demand the impossible autonomy of the consolidated bourgeois subject. In the global era, that kind of historical effectivity—of the individual personality—becomes more and more fantastic (and ideological) in relation to a reality that is smoothed out into more and more massive protagonists. At a certain point, expelled from this shell by the pressure of a flattening history, we will have to find some other kind of person—some collaborative and collective person—that we can come to be against what is now given as identity. This is a new kind of being in one's body, which is immediately the complex and whole body of human struggle. Individual moments of this struggle no longer appear as particular manifestations of a universal problematic but instead as the same whole problematic itself, at each of its infinite edges. The task for critical pedagogy is less to uncover the essence of which the particular is only the form of appearance and more to produce a complex reading of this new universal-particular from the multiple informations that are given by the different standpoints taken on it. "Bring Najaf to New York," as Klein (2004) puts it, with reference to the demands of the peace movement.

Critical pedagogy in the present needs to demythologize not only the order of reality as given but also "transformative" education as it is officially offered. In dominant as well as progressive approaches, students are "empowered" into the cells of their own "agency"; this agency becomes a possession or property to deploy. In this way, in popular approaches to critical education, identities are "empowered" into their own recuperation by the logic of capital (McLaren 2003). This is true of many progressive versions of teacher education as well, which end up reabsorbing students into the "cognitive passivity" (Kincheloe 1993) that characterizes mainstream approaches. In this context, critical pedagogy needs to pursue a Brechtian strategy of estrangement[4] in order not to *improve* the selves that students get to inhabit but to begin to make those selves, as given by power, actually visible and escapable.

Humanism should inform this critical pedagogy of the global, not in an abstract sense but instead as a set of embodied relationships (Darder 2002). Within these relationships, pedagogy should learn to evade the official positions of teacher and student, as the processes of teaching and learning point continually to a sense of humanity that overflows those positions. In other words, the Freirean positions of "teacher-student" and "student-teacher" get further smeared out so that the profile of pedagogical authority is lowered without being removed. In this way, pedagogy is less an unveiling and more the production of a fabric of opposition. This fabric links the individual positions in it into a material totality, effective beyond the sites that comprise it, without thereby constituting a simple unity. Praxis moves outward to fix, from an initial solidarity, the solidity of organization—against the correspondingly extended antagonist that can then be recognized as the body of power and capital. This fundamental intuition of *against*, which is more and more stripped bare, remains to orient the collective building of a liberatory movement and pedagogy. In this way, the paradoxical emphasis for pedagogy in the present becomes *what it learns*. A truer collaboration works to find a newer, less suspected knowledge. With the sanction of authority lighter on the body of the truth, this finding is that much more of a making.

Conclusion

Is it possible to imagine an emerging culture and pedagogy of opposition that would be equal to the global scale of capital itself and yet rooted in the materiality of human experience? Does the global necessarily transcend any particular experience, or is there a planetary particularity that might be called human and that remains tied to a particular geography, namely this earth? As paradoxical or difficult as it appears, I believe that this is the task of critical theory and pedagogy in the present: to participate in articulating an oppositional planetary identity that draws its strength from the histories of resistance that the vast majority who live and have lived have shared. If this does not make of the human a simple abstraction, this is because the broad particularity that this culture would be built on is the particularity of the global majority, as against the elite. The determined content of this culture must be the life of struggle, in all its colors and shapes—blending the cities and the *selva*, the soil and the concrete. Its unprecedented variegation, velocity, and complexity must even begin to surpass the limits of the hybrid and to become something else—a new language and color, never before seen or heard. What would make this culture authentic is what has always made culture authentic in modernity—a participation in resistance; a refusal of exploitation, occupation, colonization, and recolonization; and the construction of a form of being that looks beyond bondage. This was Frantz Fanon's (1963) principle for culture in the context of national liberation struggles against Europe's fraying empires: not a repetition of precolonial forms but an expression of the contemporary aspirations of the people.

In the present, the challenge is to overcome a simple backward-looking allegiance to the old places and to find a language for the *new local*, as it is increasingly materialized by the changing conditions of life: namely, the global itself. A new global language of resistance must start from the ground up, from the collision of living vernaculars and actual itineraries of human experience rather than from some pregiven image. The first stages of this transformation can be seen in the transnationalism of new cultural idioms emerging from the "global cities" in the street-level cosmopolitanism of popular music and popular *intifadas*. The next stages cannot be foreseen but will have to involve an actual slipping of the boundaries that are now only crossed over. In this regard, if the social forces that assault and reorganize people's lives are now properly global, critical pedagogy must also move towards an oppositional transnationalism that supports the resistant expressions of young people, is in solidarity with radical educators everywhere, and forges a path with others toward new forms of social life. The democratic and resistant identification that waits to be created in this connection is not simply a recognition of oneself in the struggle of *this* group, but rather a recognition of oneself in *all* struggles. As this spirit finds its material form and expression, we will better understand what a global identification can be, and what it means to renew a humanist commitment even as the content of the human is changing. To teach is to keep the paths open to this place rather than to decide what it must be.

Notes

Chapter 1

1. Paulo Freire, *Pedagogy of the Oppressed*, trans. M. B. Ramos (New York: Continuum, 1997), 25–26. This work will be cited as *PO* in this chapter for all subsequent references.
2. While Giroux (1993) has usefully suggested a reading of Freire as a postcolonial intellectual, my focus is on some crucial tensions *between* Freire's thought and the work of leading theorists of postcolonial societies. It should be pointed out that Freire was disciplined about continuing to critically rethink his own work; see in particular Freire (1999). However, he could not fully anticipate the new challenges with which contemporary changes in the nature of power and oppression (as well as efforts to theorize these conditions) would confront liberatory education.
3. Fischman (2001) contrasts this sense of hope with the popular and idealistic version (especially in education) which removes it from the context of concrete and historical struggle.
4. See Adorno's *Negative Dialectics* (1995). The tensions between the thought of the Frankfurt School and the orthodox Marxist perspective as represented in critical pedagogy are described in detail by Ilan Gur-Ze'ev (2005). See also the seminal discussion in Giroux (2001).
5. See Chatterjee (1986) and Guha (1988) for an analysis of this phenomenon with regard to the discourses of Indian nationalism and nationalist historiography.
6. With some notable exceptions; above all, see Lipman (2004).
7. In this connection, it is important to point out the recent partial return to an emphasis on class struggle within Freirean approaches, most importantly in the work of Peter McLaren (e.g., 2000, 2005).
8. See for example Brecht's *Mother Courage* (1955), which is an unparalleled exposition of the ambiguities of both bourgeois and revolutionary virtue in the face of the concrete contradictions of war and exploitation.

Chapter 3

1. The broader problem here is that Marxist analysis must be wary of attributing to itself the "uniquely valid standards of rationality, objectivity, [and] method" that Harding (1998, 165) describes as being commonly and erroneously associated with modern European science more generally.
2. Here an intuition of the *formal* unevenness of the social as "structured complex unity" (Althusser 1996, 199) comes together with a registering of the *experiential* priority of "secondary" contradictions like racism, sexism, or xenophobia; in a theoretical project oriented toward struggle, this experiential priority must translate in part into analytical priority.
3. See Hennessy (1993) for an extended discussion of the logical and political contradictions within feminist standpoint theory.
4. In this connection, see Street (2005) for a thorough critique of recent "adequacy"-based efforts in state courts to challenge educational disparities; in the effort to win short-term and minimal gains, these efforts lose track of the deeper logic of racism in education, and the necessity of struggle against it.

Chapter 4

1. This position has been most influentially articulated by David Harvey (2003, 2005), whose analysis (which I describe in the following section) inspires my return to Marx's own account in this chapter.
2. The connections between the current conjuncture in the global economy and trends in education have been investigated by a number of writers. Saltman (2007) very usefully applies Harvey's account to the context of the privatization of schools. In addition, see for example Lipman (2004), McLaren (2003), and Torres (2002).
3. Marx's use of this term is initially partly ironic, since he adopts the idea of primitive accumulation (an original accumulation of wealth that precedes capitalist accumulation and makes possible its development) from bourgeois political economists, in whose mythology it was a result of the special diligence and frugality of exemplary individuals (see Marx 1867/1976, 873–74). Marx's purpose is to reveal the real and ignominious history of this period, and its role in the larger dialectic of class struggle.
4. Marx's essential description of the historical passage I discuss in this chapter is given in "So-Called Primitive Accumulation," in *Capital*, vol. 1, trans. B. Fowkes (London: Penguin, 1976), 871–940.
5. Edward Said (1993) has shown how the experience of imperialism is inscribed, overtly or covertly, at the most rarefied levels of European

literature. Marx anticipates this analysis in tracing the dependence of the economic and cultural development of the West on the conquest of the non-European world.
6. This new opportunism on the part of capital, which is prepared to seize on all disasters whether social or natural, has been described by Naomi Klein (2007) in more general terms, and with reference to catastrophes across the globe, as "disaster capitalism."
7. See this firm's own description of the "learning environment" in its schools at http://www.edisonschools.com.
8. Later, Foucault (1977) will give a less economically motivated account of this process of formation of modern disciplinarity, subjectivity, and deviance. In both Marx and Foucault, however, modern criminality is a social construction tied to deep changes in the modes of production and regulation of society.
9. From this perspective, as Sandy Grande (2007) argues, struggle in education must be both a decolonizing and a critical project.
10. These considerations suggest an addendum to Marx's own account. The invention of capitalism as a mode of social and economic production, and the fantastic violence on a global scale that made possible its expansion, must be understood as well in terms of a spiritual violation and colonization of human being, and as a remaking of its soul in terms compatible with the reificatory and assimilationist logic of the "market."

Chapter 5

1. In exposing this violence at the level of the word, Derrida (1974) in particular anticipated in important ways later critiques focused on culture and politics, especially in postcolonial studies.
2. The recent work of Hardt and Negri (2000, 2004) is an important starting point for this project. However, as I discuss in Chapter 6, in their conception of the democratic subject ("multitude") that materializes the commonality of oppositional movements globally, they abstract from the actual differences that allow for the complex and territorialized sense of hybridity I suggest here.

Chapter 6

1. In this connection, see also Mies (1998).
2. Freire defines limit-situations as constituted by systems of social relationships that "imply the existence of persons who are directly or indirectly

served by these situations, and of those who are negated and curbed by them" (1997, 83).
3. See LaDuke (2005) for an important account of contemporary indigenous struggles to reclaim land and cultural property in North America.
4. See Balakrishnan (2003) for a useful collection of commentaries by leading left critics on the central arguments of Hardt and Negri in *Empire*.
5. I take *terran* from the Latin *terra*, which combines the meanings of (1) ground or soil, and (2) land (as opposed to the heavens). In the modern era, these initial meanings are articulated naturally to the larger sense of Earth as planet; it is the *combination* of these senses that I mean to evoke.
6. This is where, as a class project, the terran would break with left-liberal efforts simply to deepen democracy as process or procedure, even when those efforts are sensitive to the challenges posed by the scale of the global (see for instance Benhabib [2002]).

Chapter 7

1. Some portions of this chapter are based on my essay, "Toward a Critical Pedagogy of the Global," in *Critical Pedagogy in Uncertain Times*, ed. Sheila Macrine, Peter McLaren, and Dave Hill (New York: Palgrave Macmillan, in press).
2. "It is posited within the concept of capital that the objective conditions of labor—and these are its own product—take on a *personality* towards it, or, what is the same, that they are posited as the property of a personality alien to the worker. The concept of capital contains the capitalist" (Marx 1973, 512).
3. Here I propose an expansion of the meanings of the principles of accumulation and dispossession, in the context of contemporary globalization, that is analogous to the elaboration I suggested at the end of Chapter 4 with reference to the notion of "primitive accumulation."
4. Brecht believed that theater should actively reveal its own artifice and constructedness as a way of inviting the audience to consider how the reality it represents might be made different.

References

Adam, H. M. 1999. Fanon as a democratic theorist. In *Rethinking Fanon: The continuing dialogue*, ed. N. C. Gibson, 119–40. Amherst, NY: Humanity Books.
Adorno, T. W. 1995. *Negative dialectics*. New York: Continuum.
Alamillo, L., D. Palmer, C. Viramontes, and E. E. Garcia. 2005. California's English-only policies: An analysis of initial effects. In *Leaving children behind: How "Texas-style" accountability fails Latino youth*, ed. A. Valenzuela, 201–24. Albany, NY: State University of New York Press.
Althusser, L. 1971. Ideology and ideological state apparatuses (Notes towards an investigation). In *Lenin and philosophy and other essays*, trans. B. Brewster, 85–126. New York: Monthly Review Press.
———. 1996. *For Marx*, trans. B. Brewster. London: Verso.
Amin, S. 1997. *Capitalism in the age of globalization: The management of contemporary society*. London: Zed Books.
Anagnostopoulos, D. 2006. "Real students" and "true demotes": Ending social promotion and the moral ordering of urban high schools. *American Educational Research Journal* 43, no. 1:5–42.
Anderson, B. 1991. *Imagined communities: Reflections on the origin and spread of nationalism*. London: Verso.
Anyon, J. 2005. *Radical possibilities: Public policy, urban education, and a new social movement*. New York: Routledge.
Anzaldúa, G. 1987. *Borderlands/la frontera: The new mestiza*. San Francisco: Aunt Lute.
Appadurai, A. 2006. *Fear of small numbers: An essay on the geography of anger*. Durham, NC: Duke University Press.
Appiah, K. A. 2006. *Cosmopolitanism: Ethics in a world of strangers*. New York: W. W. Norton & Company.
Apple, M. W. 1990. *Ideology and curriculum*. 2nd ed. New York: Routledge.
———. 2001. *Educating the "right" way: Markets, standards, God, and inequality*. New York: RoutledgeFalmer.
Arnot, M. 1982. Male hegemony, social class, and women's education. *Journal of Education* 164, no. 1:64–89.

Aronowitz, S. 2000. The New World Order (They mean it). *The Nation*, July 17.
———. 2006. *Left turn: Forging a new political future*. Boulder, CO: Paradigm Publishers.
Bakhtin, M. M. 1981. *The dialogic imagination*, trans. C. Emerson and M. Holquist. Austin, TX: University of Texas Press.
Balakrishnan, G. (Ed.). 2003. *Debating Empire*. London: Verso.
Bauman, Z. 2000. *Liquid Modernity*. Cambridge, England: Polity.
Benhabib, S. 2002. *The claims of culture: Equality and diversity in the global era*. Princeton, NJ: Princeton University Press.
Berlowitz, M. J., and N. A. Long. 2003. The proliferation of JROTC: Educational reform or militarization? In *Education as enforcement: The militarization and corporatization of schools*, ed. K. J. Saltman and D. A. Gabbard, 163–74. New York: Routledge.
Bhabha, H. K. 1994. *The location of culture*. London: Routledge.
Blauner, R. 1972. *Racial oppression in America*. New York: Harper and Row.
Bond, P. 2006. *Talk left, walk right: South Africa's frustrated global reforms*. Kwa-Zulu Natal, South Africa: University of KwaZulu-Natal Press.
Bowles, S., and H. Gintis. 1976. *Schooling in capitalist America: Educational reform and the contradictions of economic life*. New York: Basic Books.
Brecht, B. 1955. *Mother Courage and her children*, trans. E. Bentley. New York: Grove Weidenfeld.
Brenner, R. 2002. *The boom and the bubble: The U.S. in the world economy*. New York: Verso.
Buber, M. 1970. *I and thou*, trans. W. Kaufmann. New York: Charles Scribner's Sons.
Butler, J. 1993. *Bodies that matter: On the discursive limits of sex*. New York: Routledge.
Castillo, A. 1994. *Massacre of the dreamers: Essays on Xicanisma*. New York: Penguin Books.
Chakrabarty, D. 2000. *Provincializing Europe: Postcolonial thought and historical difference*. Princeton, NJ: Princeton University Press.
Chatterjee. 1986. *Nationalist thought and the colonial world: A derivative discourse*. Minneapolis, MN: University of Minnesota Press.
Chubb, J. E., and T. M. Moe. 1990. *Politics, markets, and America's schools*. Washington, DC: The Brookings Institution.
Cochran-Smith, M. 1991. Learning to teach against the grain. *Harvard Educational Review* 61, no. 3:279–310.
Collins, P. H. 2000. *Black feminist thought: Knowledge, consciousness, and the politics of empowerment*. New York: Routledge.

Cox, O. C. 2000. *Race: A study in social dynamics*. New York: Monthly Review Press.

Currie, G., and D. Knights. 2003. Reflecting on a critical pedagogy in MBA education. *Management Learning* 34, no. 1:27–49.

Darder, A. 1991. *Culture and power in the classroom: A critical foundation for bicultural education*. Westport, CT: Bergin & Garvey.

———. 2002. *Reinventing Paulo Freire: A pedagogy of love*. Boulder, CO: Westview Press.

Darder, A., and R. D. Torres. 2003. Shattering the "race" lens: Toward a critical theory of racism. In *The critical pedagogy reader*, ed. A. Darder, M. Baltodano, and R. D. Torres, 245–61. New York: RoutledgeFalmer.

Dead Prez (C. Galvin, L. Alford, V. Williams, A. Mair). 2000. "Behind enemy lines," on *Let's Get Free*. Loud Records.

De Lissovoy, N. 2004. Affirmation, ambivalence, autonomy: Reading the subaltern subject in postcolonial historiography and critical pedagogy. *Journal of Postcolonial Education* 3, no. 1:5–23.

———. 2007. Frantz Fanon and a materialist critical pedagogy. In *Critical Pedagogy: Where Are We Now?* ed. P. McLaren and J. L. Kincheloe, 355–70. New York: Peter Lang.

———. 2007. History, histories, or historicity? The time of educational liberation in the age of empire. *Review of Education, Pedagogy, and Cultural Studies* 29, no. 5:441–60.

———. 2008. Conceptualizing oppression in educational theory: Toward a compound standpoint. *Cultural Studies ↔ Critical Methodologies* 8, no. 1:82–105.

De Lissovoy, N., and P. McLaren. 2003. Educational "accountability" and the violence of capital: A Marxian reading. *Journal of Education Policy* 18, no. 2:131–43.

———. 2005. Toward a contemporary philosophy of praxis. In *Radical relevance: Toward a scholarship of the whole left*, ed. L. Gray-Rosendale and S. Rosendale, 160–82. Albany, NY: State University of New York Press.

———. 2006. Ghosts in the procedure: Notes on teaching and subjectivity in a new era. In *The practical critical educator*, ed. K. Cooper and R. E. White, 151–63. Dordrecht: Springer.

Deloria, V., Jr., and C. Lytle. 1984. *The nations within: The past and future of American Indian sovereignty*. New York: Pantheon.

Delpit, L. 1995. *Other people's children: Cultural conflict in the classroom*. New York: The New Press.

Derrida, J. 1974. *Of grammatology*, trans. G. C. Spivak. Baltimore, MD: Johns Hopkins University Press.

Desai, A. 2002. *We are the poors: Community struggles in post-apartheid South Africa*. New York: Monthly Review Press.

Devine, J. 1996. *Maximum security: The culture of violence in inner-city schools*. Chicago: University of Chicago Press.

Dewey, J. 1944. *Democracy and education*. New York: The Free Press.

Dohrn, B. 2001. "Look out kid/ it's something you did": Zero tolerance for children. In *Zero tolerance: Resisting the drive for punishment in our schools*, ed. W. Ayers, B. Dohrn, and R. Ayers, 89–113. New York: The New Press.

Eagleton, T. 1991. *Ideology: An introduction*. London: Verso.

Ellsworth, E. 1997. *Teaching positions: Difference, pedagogy, and the power of address*. New York: Teachers College Press.

Fanon, F. 1963. *The wretched of the earth*, trans. C. Farrington. New York: Grove Press.

———. 1965. *A dying colonialism*, trans. H. Chevalier. New York: Grove Press.

———. 1967a. *Black skin, white masks*, trans. C. L. Markmann. New York: Grove Press.

———. 1967b. *Toward the African revolution*, trans. H. Chevalier. New York: Grove Press.

Fischman, G. E. 2001. Teachers, globalization, and hope: Beyond the narrative of redemption. *Comparative Education Review* 45, no. 3:412–18.

Flax, J. 1990. *Thinking fragments: Psychoanalysis, feminism, and postmodernism in the contemporary West*. Berkeley, CA: University of California Press.

Foucault, M. 1977. *Discipline and punish*, trans. A. Sheridan. New York: Pantheon Books.

———. 1980. *Power/knowledge: Selected interviews and other writings 1972–1977*. New York: Pantheon.

Freire, P. 1978. *Pedagogy in process: The letters to Guinea-Bissau*. New York: Seabury Press.

———. 1997. *Pedagogy of the oppressed*, trans. M. B. Ramos. New York: Continuum.

———. 1998. *Teachers as cultural workers: Letters to those who dare teach*, trans. D. Macedo, D. Koike and A. Oliveira. Boulder, CO: Westview Press.

———. 1999. *Pedagogy of hope*, trans. R. R. Barr. New York: Continuum.

Freire, P., and D. Macedo. 1987. *Literacy: Reading the word and the world*. South Hadley, MA: Bergin and Garvey.

Friedman, T. L. 2005. *The world is flat: A brief history of the twenty-first century*. New York: Farrar, Straus and Giroux.

Furumoto, R. 2005. No poor child left unrecruited: How NCLB codifies and perpetuates urban school militarism. *Equity and Excellence in Education*, 38, no. 3:200–10.

Gadotti, M. 1994. *Reading Paulo Freire*, trans. J. Milton. Albany, NY: State University of New York Press.

Gay, G. 2000. *Culturally responsive teaching: Theory, research, and practice*. New York: Teachers College Press.

Gibson, N. C. 1999. Radical mutations: Fanon's untidy dialectic of history. In *Rethinking Fanon: The continuing dialogue*, ed. N. Gibson, 408–46. Amherst, NY: Humanity Books.

Gilroy, P. 1993. *The Black Atlantic: Modernity and double consciousness*. Cambridge, MA: Harvard University Press.

Giroux, H. A. 1988. *Schooling and the struggle for public life: Critical pedagogy in the modern age*. Minneapolis, MN: University of Minnesota Press.

———.1992. *Border crossings: Cultural workers and the politics of education*. New York: Routledge.

———. 1993. Paulo Freire and the politics of postcolonialism. In *Paulo Freire: A critical encounter*, ed. P. McLaren and P. Leonard, 177–88. New York: Routledge.

———. 2000. Postmodern education and disposable youth. In *Revolutionary pedagogies: Cultural politics, instituting education, and the discourse of theory*, ed. P. P. Trifonas, 174–95. New York: Routledge.

———. 2001. *Theory and resistance in education*. Westport, CT: Bergin & Garvey.

———. 2003. *The abandoned generation: Democracy beyond the culture of fear*. New York: Palgrave Macmillan.

Gluckman, A. 2002. Testing . . . testing . . . one, two, three: The commercial side of the standardized-testing boom. *Dollars & Sense* 239. http://dollarsandsense.org/archives/2002/0102gluckman.html.

Gowan, P. 2003. U.S. Hegemony Today. *Monthly Review* 55, no. 3:30–50.

Grande, S. 2000. American Indian identity and intellectualism: The quest for a new Red pedagogy. *Qualitative Studies in Education* 13, no. 4:343–59.

———. 2007. Red Lake Woebegone: Pedagogy, decolonization, and the critical project. In *Critical pedagogy: Where are we now?* ed. P. McLaren and J. L. Kincheloe, 315–36. New York: Peter Lang.

Guha, R. 1988. On some aspects of the historiography of colonial India. In *Selected Subaltern Studies*, ed. R. Guha and G. C. Spivak, 37–44. Oxford: Oxford University Press.

Gur-Ze'ev, I. 2005. Adorno and Horkheimer: Diasporic philosophy, negative theology, and counter-education. *Educational Theory* 55, no. 3:343–65.

Gutiérrez, K., J. Asato, and P. Baquedano-López. 2000. "English for the Children": The new literacy of the old world order, language policy and educational reform. *Bilingual Research Journal* 24, nos. 1, 2, and 3:87–107.

Gutiérrez, K., J. Asato, M. Santos, and N. Gotanda. 2002. Backlash pedagogy: Language and culture and the politics of reform. *Review of Education, Pedagogy, and Cultural Studies* 24:335–51.

Gutiérrez, K., P. Baquedano-López, and C. Tejeda. 1999. Rethinking diversity: Hybridity and hybrid language practices in the third space. *Mind, Culture, and Activity* 6, no. 4:286–303.

Gutiérrez, K., B. Rymes, and J. Larson. 1995. Script, counterscript, and underlife in the classroom: James Brown versus *Brown v. Board of Education*. *Harvard Educational Review* 65, no. 3:445–71.

Habermas, J. 1984. *Reason and the Rationalization of Society.* Vol. 1 of *The Theory of Communicative Action*, trans. T. McCarthy. Boston, MA: Beacon Press.

———. 1987. *Lifeworld and System: A Critique of Functionalist Reason.* Vol. 2 of *The Theory of Communicative Action*, trans. T. McCarthy. Boston, MA: Beacon Press.

Hall, S. 1986. The problem of ideology: Marxism without guarantees. *Journal of Communication Inquiry* 10, no. 2:28–44.

Harding, S. 1993. Rethinking standpoint epistemology: What is "strong objectivity"? In *Feminist Epistemologies*, ed. L. Alcoff and E. Potter, 49–82. New York: Routledge.

———. 1998. *Is science multicultural? Postcolonialisms, feminisms, and epistemologies.* Bloomington, IN: Indiana University Press.

Hardt, M., and A. Negri. 2000. *Empire.* Cambridge, MA: Harvard University Press.

———. 2004. *Multitude: War and democracy in the age of empire.* New York: Penguin Press.

Harris, C. I. 1995. Whiteness as property. In *Critical race theory: The key writings that formed the movement*, ed. K. Crenshaw, N. Gotanda, G. Peller, and K. Thomas, 276–91. New York: New York Press.

Hartsock, N. 1983. The feminist standpoint: Developing the ground for a specifically feminist historical materialism. In *Discovering reality: Feminist perspectives on epistemology, metaphysics, methodology, and philosophy of science*, ed. S. Harding and M. B. Hintikka, 283–310. Dordrecht: D. Reidel.

Harvey, D. 2003. *The new imperialism.* Oxford: Oxford University Press.

———. 2005. *A brief history of neoliberalism.* Oxford: Oxford University Press.

Haymes, S. N. 1995. *Race, culture, and the city: A pedagogy for black urban struggle.* Albany, NY: State University of New York Press.

Hennessy, R. 1993. *Materialist feminism and the politics of discourse.* New York: Routledge.

Holloway, J. 2002. *Change the world without taking power*. London: Pluto Press.

hooks, b. 1994. *Teaching to transgress: Education as the practice of freedom*. New York: Routledge.

Horkheimer, M., and T. W. Adorno. 2002. *Dialectic of enlightenment*, trans. E. Jephcott. Stanford, CA: Stanford University Press.

Howard, T. C. 2001. Powerful pedagogy for African American students: A case of four teachers. *Urban Education* 36, no. 2:179–202.

Jackson, G. 1970. *Soledad brother*. New York: Coward-McCann.

Jaggar, A. 1989. Love and knowledge: Emotion in feminist epistemology. *Inquiry* 32:151–72.

James, C. L. R. 1992. Dialectical materialism and the fate of humanity. In *The C. L. R. James Reader*, ed. A. Grimshaw, 153–81. Cambridge, MA: Blackwell.

Jameson, F. 1991. *Postmodernism, or, the cultural logic of late capitalism*. Durham, NC: Duke University Press.

Kaomea, J. 2003. Reading erasures and making the familiar strange: Defamiliarizing methods for research in formerly colonized and historically oppressed communities. *Educational Researcher* 32, no. 2:14–25.

Kelley, R. D. G. 1997. *Yo' mama's disfunktional: Fighting the culture wars in urban America*. Boston, MA: Beacon Press.

Kincheloe, J. L. 1993. *Toward a critical politics of teacher thinking: Mapping the postmodern*. Westport, CT: Bergin & Garvey.

———. 2007. Critical pedagogy in the twenty-first century: Evolution for survival. In *Critical Pedagogy: Where Are We Now?* ed. P. McLaren and J. L. Kincheloe, 9–42. New York: Peter Lang.

Kincheloe, J. L., and S. R. Steinberg. 1997. *Changing multiculturalism*. Bristol, PA: Open University Press.

Klein, N. 2004. Bring Najaf to New York. *The Nation* 279 (September 13): 22.

———. 2007. *The shock doctrine: The rise of disaster capitalism*. New York: Metropolitan Books.

Kovel, J. 2002. *The enemy of nature: The end of capitalism or the end of the world?* New York: Zed Books.

Kozol, J. 2005. *The shame of the nation: The restoration of apartheid schooling in America*. New York: Crown Publishers.

———. 2007. The big enchilada. *Harper's* 315 (August): 7–9.

Kumashiro, K. K. 2002. Against repetition: Addressing resistance to anti-oppressive change in the practices of learning, teaching, supervising, and researching. *Harvard Educational Review* 72, no. 1:67–92.

Laclau, E., and C. Mouffe. 1985. *Hegemony and socialist strategy: Towards a radical democratic politics.* London: Verso.

Ladson-Billings, G. 1992. Reading between the lines and beyond the pages: A culturally relevant approach to literacy teaching. *Theory into Practice* 34, no. 4:312–20.

———. 1994. *The dreamkeepers: Successful teachers of African American children.* San Francisco, CA: Jossey-Bass.

———. 2001. *Crossing over to Canaan: The journey of new teachers in diverse classrooms.* San Francisco, CA: Jossey-Bass.

———. 2004. Landing on the wrong note: The price we paid for *Brown. Educational Researcher* 33, no. 7:3–13.

LaDuke, W. 2005. *Recovering the sacred: The power of naming and claiming.* Cambridge, MA: South End Press.

Lather, P. 1998. Critical pedagogy and its complicities: A praxis of stuck places. *Educational Theory* 48, no. 4:487–97.

Lee, C. D., and P. Smagorinsky. 2000. *Vygotskian perspectives on literacy research: Constructing meaning through collaborative inquiry.* Cambridge, England: Cambridge University Press.

Leistyna, P. A. 2004. *Revolutionary possibilities.* Paper presented at the American Educational Research Association, San Diego, CA.

Leonardo, Z. 2002. The souls of white folk: Critical pedagogy, whiteness studies, and globalization discourse. *Race, Ethnicity and Education* 5, no. 1:29–50.

———. 2005. The Color of supremacy: Beyond the discourse of "white privilege." In *Critical Pedagogy and Race*, ed. Z. Leonardo, 37–52. Malden, MA: Blackwell Publishing.

Lipman, P. 2004. *High stakes education: Inequality, globalization, and urban school reform.* New York: RoutledgeFalmer.

Lorde, A. 1978. *The black unicorn.* New York: Norton.

Lukács, G. 1971. *History and class consciousness*, trans. R. Livingstone. Cambridge, MA: MIT Press.

Luxemburg, R. 2004. Organizational questions of Russian social democracy. In *The Rosa Luxemburg Reader*, ed. P. Hudis and K. B. Anderson, 248–65. New York: Monthly Review Press.

Lyons, S. R. 2005. The left side of the circle: American Indians and progressive politics. In *Radical relevance: Toward a scholarship of the whole left*, ed. L. Gray-Rosendale and S. Rosendale, 69–84. Albany, NY: State University of New York Press.

Lyons, W., and J. Drew. 2006. *Punishing schools: Fear and citizenship in American public education.* Ann Arbor, MI: University of Michigan Press.

Macedo, D. 2000. The colonialism of the English only movement. *Educational Researcher* 29, no. 3:15–24.

Marx, K. 1867/1976. *Capital*, vol. 1, trans. B. Fowkes. London: Penguin Books.

———. 1964. *The economic and philosophic manuscripts of 1844*, trans. M. Milligan. New York: International Publishers.

———. 1973. *Grundrisse*, trans. M. Nicolaus. London: Penguin Books.

Marx, K., and F. Engels. 1848/1967. *The Communist manifesto*, trans. S. Moore. London: Penguin Books.

———. 1970. *The German ideology*. New York: International Publishers.

McChesney, R. W. 1999. *Rich media, poor democracy: Communication politics in dubious times*. Urbana, IL: University of Illinois Press.

McClintock, A. 1997. "No longer in a future heaven": Gender, race and nationalism. In *Dangerous Liaisons: Gender, Nation, and Postcolonial Perspectives*, ed. A. McClintock, A. Mufti, and E. Shohat, 89–112. Minneapolis, MN: University of Minnesota Press.

McLaren, P. 1997. *Revolutionary multiculturalism: Pedagogies of dissent for the new millenium*. Boulder, CO: Westview Press.

———. 2000. *Che Guevara, Paulo Freire, and the pedagogy of revolution*. Lanham, MD: Rowman and Littlefield.

———. 2001. Bricklayers and bricoleurs: A Marxist addendum. *Qualitative Inquiry* 7, no. 6:700–705.

———. 2003. Critical pedagogy and class struggle in the age of neoliberal globalization: Notes from history's underside. *Democracy & Nature* 9, no. 1:65–90.

———. 2005. *Capitalists and conquerors: A critical pedagogy against empire*. Lanham, MD: Rowman & Littlefield.

McLaren, P., and R. Farahmandpur. 2005. *Teaching against global capitalism and the new imperialism: A critical pedagogy*. Lanham, MD: Rowman & Littlefield.

McNeil, L. M. 2000. *Contradictions of school reform: Educational costs of standardized testing*. New York: Routledge.

———. 2005. Faking equity: High-stakes testing and the education of Latino youth. In *Leaving children behind: How "Texas-style" accountability fails Latino youth*, ed. A. Valenzuela, 57–111. Albany, NY: State University of New York Press.

Mertes, T. 2003. Grass-roots globalism. In *Debating Empire*, ed. G. Balakrishnan, 144–54. London: Verso.

Mészáros, I. 1995. *Beyond capital*. New York: Monthly Review Press.

Meyer, R. J. 2002. Captives of the script: Killing us softly with phonics. *Language Arts* 79, no. 6:452–61.

Mies, M. 1998. *Patriarchy and accumulation on a world scale: Women in the international division of labour.* London: Zed Books.

Mignolo, W. D. 2004. Globalization, civilization processes, and the relocation of languages and cultures. In *The Cultures of Globalization,* ed. F. Jameson and M. Miyoshi, 32–53. Durham, NC: Duke University Press.

Miner, B. 2004/2005. Keeping schools public: Testing companies mine for gold. *Rethinking Schools* 19. http://www.rethinkingschools.org/archive/19_02/test192.shtml.

Mohanty, C. T. 2003. *Feminism without borders: Decolonizing theory, practicing solidarity.* Durham, NC: Duke University Press.

Molnar, A. 2005. *School commercialism: From democratic ideal to market commodity.* New York: Routledge.

Negri, A. 1984. *Marx beyond Marx: Lessons on the* Grundrisse, trans. H. Cleaver, M. Ryan, and M. Viano. South Hadley, MA: Bergin & Garvey.

New London Group 1996. A pedagogy of multiliteracies: Designing social futures. *Harvard Educational Review* 66, no. 1:60–92.

Nieto, S. 1998. Fact and fiction: Stories of Puerto Ricans in U.S. schools. *Harvard Educational Review* 68, no. 2:133–63.

———. 1999. *The light in their eyes: Creating multicultural learning communities.* New York: Teachers College Press.

———. 2002. *Language, culture, and teaching: Critical perspectives for a new century.* Mahwah, NJ: Lawrence Erlbaum.

———. 2004. *Affirming diversity: The sociopolitical context of multicultural education.* 4th ed. Boston, MA: Pearson Education.

Oakes, J., and M. Saunders. 2004. Education's most basic tools: Access to textbooks and instructional materials in California's public schools. *Teachers College Record,* 106, no. 10:1967–1988.

Omi, M., and H. Winant. 1994. *Racial formation in the United States: From the 1960s to the 1990s.* New York: Routledge.

Orellana, M. F. 2001. The work kids do: Mexican and Central American immigrant children's contributions to households and schools in California. *Harvard Educational Review* 71, no. 3:366–89.

Pollock, S., H. K. Bhabha, C. A. Breckenridge, and D. Chakrabarty. 2000. Cosmopolitanisms. *Public Culture* 12, no. 3:577–89.

Przeworski, A. 1986. *Capitalism and social democracy.* Cambridge: Cambridge University Press.

Reid, D. K., and M. G. Knight. 2006. Disability justifies exclusion of minority students: A critical history grounded in disability studies. *Educational Researcher* 35, no. 6:18–23.

Robinson, W. I. 1996. Globalisation: Nine theses on our epoch. *Race & Class* 38, no. 2:13–31.

Said, E. W. 1993. *Culture and Imperialism*. New York: Vintage Books.
Saltman, K. J. 2005. *The Edison Schools: Corporate schooling and the assault on public education*. New York: Routledge.
———. 2007. *Capitalizing on disaster: Taking and breaking public schools*. Boulder, CO: Paradigm Publishers.
Saltman, K. J., and D. A. Gabbard, eds. 2003. *Education as enforcement: The militarization and corporatization of schools*. New York: RoutledgeFalmer.
Sandoval, C. 2000. *Methodology of the oppressed*. Minneapolis: University of Minnesota Press.
Sassen, S. 1998. *Globalization and its discontents: Essays on the new mobility of people and money*. New York: The New Press.
Schutz, A. 2004. Rethinking domination and resistance: Challenging postmodernism. *Educational Researcher* 33, no. 1:15–23.
Scribner, S. 1984. The practice of literacy: Where mind and society meet. *Annals of the New York Academy of Sciences*, no. 433:5–19.
Shiva, V. 2005. *Earth democracy: Justice, sustainability, and peace*. Cambridge, MA: South End Press.
Sklair, L. 2002. *Globalization: Capitalism and its alternatives*. Oxford: Oxford University Press.
Sleeter, C. E. 1993. How white teachers construct race. In *Race, Identity, and Representation in Education*, ed. C. McCarthy and W. Crichlow, 157–71. New York: Routledge.
Spivak, G. C. (1988a). Can the subaltern speak? In *Marxism and the interpretation of culture*, ed. C. Nelson and L. Grossberg, 271–313. Chicago: University of Illinois Press.
———. (1988b). Subaltern studies: Deconstructing historiography. In *Selected Subaltern Studies*, ed. R. Guha and G. C. Spivak, 3–32. Oxford: Oxford University Press.
Starnes, B. A. 2000. On dark times, parallel universes, and déjà vu. *Phi Delta Kappan* 82, no. 2:108–14.
Stoskopf, A. 1999. The forgotten history of eugenics. *Rethinking Schools* 13:12–13.
Street, P. 2005. *Segregated schools: Educational apartheid in post-civil rights America*. New York: Routledge.
Torres, C. A. 2002. Globalization, education, and citizenship: Solidarity versus markets? *American Educational Research Journal* 39, no. 2:363–78.
Valenzuela, A. 2005. Accountability and the privatization agenda. In *Leaving children behind: How "Texas-style" accountability fails Latino youth*, ed. A. Valenzuela, 263–94. Albany, NY: State University of New York Press.
Villegas, A. M., and T. Lucas. 2002. *Educating culturally responsive teachers: A coherent approach*. Albany, NY: State University of New York Press.

Vinson, K. D., and E. W. Ross. 2003. Controlling images: The power of high-stakes testing. In *Education as Enforcement: The Militarization and Corporatization of Schools*, ed. K. J. Saltman and D. A. Gabbard, 241–57. New York: Routledge.

Vygotsky, L. S. 1978. *Mind in society: The development of higher psychological processes*. Cambridge, MA: Harvard University Press.

Walkerdine, V. 1990. *Schoolgirl fictions*. London: Verso.

———. 1997. *Daddy's girl: Young girls and popular culture*. Cambridge, MA: Harvard University Press.

———. 1998. *Counting girls out: Girls and mathematics*. London: Falmer Press.

Weiler, K. 1988. *Women teaching for change*. Westport, CT: Bergin & Garvey.

West, C. 1988. Marxist theory and the specificity of Afro-American oppression. In *Marxism and the interpretation of culture*, C. Nelson and L. Grossberg, 17–29. Chicago: University of Illinois Press.

———. (1993a). The new cultural politics of difference. In *Race, Identity and Representation in Education*, ed. C. McCarthy and W. Crichlow, 11–23. New York: Routledge.

———. (1993b). *Keeping Faith: Philosophy and Race in America*. New York: Routledge.

Westheimer, J., and J. Kahne. 1998. Education for action—Preparing youth for participatory democracy. In *Teaching for Social Justice*, ed. W. Ayers, J. A. Hunt, and T. Quinn, 1–20. New York: The New Press.

Willis, P. 1977. *Learning to labor: How working class kids get working class jobs*. New York: Columbia University Press.

Young, I. M. 1990. *Justice and the politics of difference*. Princeton, NJ: Princeton University Press.

Žižek, S. 2002. *Welcome to the desert of the real*. London: Verso.

Index

The letter *n* following a page number indicates a note on that page.

accountability initiatives, 89, 92, 118, 164
accumulation, 50, 59–63, 106, 136, 137–139, 142; empire and, 140; technology of solidarity and, 163
activists, 66
Adam, Hussein, 45
Adorno, Theodor W., 14, 16, 24
Affirming Diversity (Nieto), 57
African-American youth, 87, 109–10
African National Congress, 133, 137
agency, 24, 65, 114; of students, 119, 176
Algeria, 45
alienation, 18, 26, 94–98, 164, 165
Amin, Samir, 167
Anagnostopoulos, D., 88
annexation, 141
anticolonial struggle, 15, 72
antiglobalization movement, 142
antioppressive education, 157
antiprivatization movements, 140
antiwar protests, 166; global, 140, 142
Anyon, J., 91
Anzaldúa, Gloria, 121
apartheid, 90, 100

Appadurai, Arjun, 151, 161
Apple, M., 20
art, 159
assistencialism, 137
authority, 44, 115; authoritarianism and, 19, 45, 74–75, 174; in contemporary schooling, 79; educational, 39–42, 176; Hardt and Negri on, 142–43; hyperauthoritarianism and, 174; leadership and, 39–42; of elites, 119–20;

Bakhtin, M. M., 113
banking education, 135–136
Bauman, Zygmunt, 17, 169–170
Benhabib, Seyla, 169
Bhabha, Homi, 43, 72, 121
biopolitics, 148, 168
biopower, 140–41
Black and brown bodies, 53, 70, 91, 101; control of, 118; students, 76, 87, 109–10. *See also* race; students of color
Black bourgeoisie, 71
Black feminist thought, 53, 69–70
Black women, 53, 70
border-crossing, 106, 107, 121, 122, 126
borders, 111

boundaries, 120
bourgeois philosophy, 47
boys, 64, 66; men, 172. *See also* gender; girls
Brazil, 135
Brecht, Bertolt, 25
Brown v. Board of Education (1954), 75, 89–90
Buber, Martin, 23
Bush, George W., 166
Butler, Judith, 35, 121

capital, 31, 130; as antagonist, 151; biopower and, 140–41; labor and, 138, 145; penetration of, 60, 96, 123, 167, 168, 170; violence of, 162
capitalism, 60, 63, 125, 181*n*6; contemporary, 132; critical pedagogy and, 39–40; critique of, 148; emergence of, 80; exploitation and, 33, 62; globality and, 158; as hybridizing force, 122; labor power and, 138, 145; naturalization of, 99; neoliberalism and, 81; overcoming, 43; overproduction and, 166–67; personality of, 165, 182*n*2; as progress, 163; reorganization of life by, 2, 81, 123–24, 143, 157, 159–60, 164–65, 169; reproduction of social life and, 64–65, 148, 149, 164, 170; revolution and, 52; school reform and, 84; social justice and, 42; struggle against, 159; success and, 38; surplus population and, 89; violence and, 33, 103; *vs.* creativity, 150
capitalist accumulation, 50, 59–63, 106, 136, 137–39, 140, 142; culture and, 131; social services and, 168; terran class and, 150. *See also* primitive accumulation
caring, 34
Chakrabarty, Dipesh, 15, 41
charter schools, 86, 92, 93–94, 97
Cheney, Dick, 166
Chicago, Illinois, 152
citizenship, 170
civil rights movement, 75
class, 90, 146, 147, 149, 171–72; accountability initiatives and, 118; colonialism and, 71; critical postmodernist pedagogy and, 111; culture and, 74; of global oppressed, 150–51; humanization and, 134; identity and, 162; labor and, 95; language of struggle in, 133; middle class, 162; neoliberalism and, 41; Nieto on, 56–57; oppressor/oppressed paradigm and, 134; politics and, 137–38; race and, 31, 39, 73; reinventing, 148; revolution and, 22, 52; in school, 75; solidarity and, 47; struggles of, 131, 133, 138, 141; students oppressed by, 68; wealth and, 31; working class, 46, 52, 68, 71, 131, 138, 141
clearings, 94, 98–101
Clinton administration, 85
codification, 44
collaboration, 24, 97–98, 113–15, 177; democracy and, 153; teacher-student, 22, 173–74
Collins, Patricia Hill, 53, 69–70
colonialism, 39, 43, 48, 120, 168; anticolonial struggle and, 15, 72; decolonization, 29, 72;

globalization and, 122; intersecting experiences of oppression in, 71; neocolonialism and, 15, 35, 75, 76–77, 135; Nieto on, 58; oppressor/oppressed paradigm and, 134; primitive accumulation and, 80; racism and, 30–33, 71; success and, 38
commercialism, 19
commonality, 106. *See also* difference
communal resources, 82–83, 84, 137, 148
communication, 45, 169; cross-cultural, 122; electronic, 175; as inherently educational, 129; multitude/empire paradigm and, 141. *See also* dialogue
communion, 22
community, 109, 153; displacement of, 96; indigenous, 58, 161
compound standpoint, 54, 70–73, 73–76. *See also* standpoint theory
conquest, 18, 122. *See also* globalization; neoliberalism
conscientization, 151, 170–71
consciousness, 11; global, 125, 154
conservatives, 168
consumerism, 97, 122, 158
contemporary schooling. *See* schools/schooling
cooperation, 141
corporations, 59–60, 158; in schools, 94–98; test industry and, 86, 87
cosmopolitanism, 150, 162
Cox, Oliver Cromwell, 58
creativity, 82, 91, 138, 172; alienation of, 165; commodification of, 167; of multitude, 140, 141–42; terran class and, 150
credentialing, 88
criminalization, 79, 84, 87, 98–101, 102, 181n8; of youth, 44
critical pedagogy, 21, 76, 172; authoritarianism and, 174; dialogue and, 46; internal/external contradictions of, 39–42; postmodernist, 108, 110–13, 114, 116; problem-posing education and, 10, 16, 22, 25–27, 135
cultural difference, 105, 107–27; critical postmodernist pedagogy and, 110–13; culturally relevant pedagogy and, 108–10; discourse and power and, 115–20; hybridity and, 120–25; pedagogy of heteroglossia and, 113–15. *See also* difference
culture, 68, 125; capitalist accumulation and, 131; class and, 74; cultural studies and, 117; as discourse, 120; dominant, 109; hegemony of, 50, 55–59, 119; imperialism and, 122; materiality and, 73, 177; oppressor/oppressed paradigm and, 134; pedagogy of heteroglossia, 114; violence and, 136. *See also* class; gender; race
curricula, 49, 72, 79; scripted, 19, 20, 55, 60, 75, 86

Dead Prez, 18
decentralization, 151–53
decolonization, 29, 72
deindustrialization, 167
Delpit, Lisa, 41
democracy, 43, 95, 120, 142, 153, 166–69; authoritarian, 45;

dismantling of, 93; in global framework, 140, 168–69; identity and, 115; pedagogy and, 44; progressive education and, 105; rethinking of, 171
dependency, 135–36, 137
Desai, Ashwin, 137, 162
desegregation, 62
detention, 88. *See also* discipline
deterritorialization, 146, 150, 161
developing countries, 133, 162
development, 14–15, 137
Dewey, John, 93, 129, 153
dialogue, 48, 130; difference and, 126; materiality and, 120; pedagogical, 26; pedagogy of heteroglossia and, 115; of politics, 22–23, 42–46. *See also* communication
diasporic identities, 120
difference, 105, 106; articulation of, 115; erasure of, 130; materiality and, 33–39; standpoint theory and, 53. *See also* cultural difference
discipline, 26, 75, 102; banking education and, 136; in charter schools, 93–94; detention as, 88; hyper-disciplinarity and, 87–89; power in education and, 118; primitive accumulation and, 83; racism and, 32; school reform and, 84; students of color and, 99
discourse, 105, 115–20
discrimination, 58, 89, 106
discursive effect, 50, 63–67
dispossession, 83, 94, 100, 133, 140, 182*n*3; accumulation by, 137–38; neoliberalism and, 138; terran class and, 150

diversity, 108, 161, 172
domination, 55, 109, 126, 134; standpoint theory and, 54
Drew, J., 88
Dying Colonialism, A (Fanon), 45

Eagleton, Terry, 116
ecological consciousness, 172
economy, 43, 59, 103, 130; global, 90; materialist analysis of, 35; of politics, 62–63, 73, 123; violence and, 136
Edison Schools, 93
education, 67–68, 126, 175; as collaborative, 22, 97–98, 113–15, 173–74, 177; difference in, 37; gender and, 64–67, 68, 74; internal/external contradictions of, 39–42; intersecting experiences of oppression and, 76–77; liberatory, 20, 172; neocolonial organization of, 72; privatization of, 60, 91–94, 139; problem-posing, 10, 16, 22, 25–27, 135; progressive, 60, 97–98, 105; as punishment, 174; racism in, 29, 32, 34–35, 38, 48, 55; radical struggle in, 61; sociopolitical context of, 50; structural adjustment and, 60, 97–98, 152; transformations in, 104, 168; transformative, 107, 176
elites, 15, 39–42, 158; authority of, 119–20; dependence on immigrants by, 160; solidarity against, 162
empire, 146, 148; accumulation and, 140; multitude and, 132, 153–54, 169; multitude/empire paradigm, 139–45; neoimperialism and,

100; reproduction of, 142; social justice movements and, 171
Empire (Hardt & Negri), 139
empowerment, 41
enclosures, 86–87, 102; in contemporary schooling, 91, 94–98
Engels, Friedrich, 36, 159, 163. *See also* Marx, Karl
England, 82
English language, 73, 85, 108
essentialism, 70
ethical commitments, 112
ethical-political conception, 146
Eurocentrism, 15–16, 27, 106
excellence in U.S. schooling, 37–39, 107
exclusion, 105
experience, 116
exploitation, 138, 142, 165; of biopower, 140–41; capitalist, 33, 62; of creativity by empire, 141; expansion of, 130; Fanon on, 30; Global Women's Strike and, 172; multitude's distance from, 143; negation and, 37; primitive accumulation and, 80; students' relationships to, 172–73; surplus population and, 89; women's labor and, 131; of workers in global South, 164; working-class struggles against, 131
expulsion, 87

failure, 37, 38
Fanon, Frantz, 29, 59, 71–72, 134, 149, 150; on dialogue, 45, 46; on difference, 34; on colonialism, 30–33, 48; liberatory movements and, 39–42; materialist analysis and, 36–38, 47;

open-ended praxis and, 43; principle for culture by, 177; standpoint theory and, 54
feminism, 62, 64, 76, 105, 121, 131; Black feminist thought and, 53, 69–70; standpoint theory and, 52–53, 70–71, 73
Foucault, Michel, 88, 141, 181n8
Freire, Paulo, 12, 44, 48, 131, 179n2; conscientization and, 151, 170–71; dialogue and, 22–23; fluid sense of oppressed and, 149; generative themes of, 173; humanization and, 9, 146; human relationships and, 153; on liberation, 61, 134, 154; Lorde and, 24–25; oppressor/oppressed paradigm, 133–39; *Pedagogy of the Oppressed*, 10, 15, 18, 138; politics and, 16–18; redemption of oppressed and, 13–14; reinvention of, 10; social psychology and, 14–16; terran class and, 150
Friedman, Thomas, 162
fundamentalism, 124

gender, 72, 121; capitalist accumulation and, 131; critical postmodernist pedagogy and, 111; globalization and, 160–61; Nieto on, 57; standpoint theory and, 52; students oppressed by, 68; Walkerdine on, 64–67, 68, 74; worker identities and, 160–61
generative themes, 136, 173
gentrification, 79, 83, 89–91, 96, 152
Gibson, Nigel, 45

girls, 64–65, 66, 68, 74; of color, 69–70. *See also* boys; gender; women
Giroux, Henry, 19, 32, 44, 111–12
global cities, 91
globality, 123, 124; alternative, 171; challenges posed by, 159; common vision of, 152; condition of, 26; individuality and, 170; as meaning of globalization, 158; possibilities of, 159–60; of terran identity, 149
globalization, 26, 91, 106, 157–78; challenges posed by, 158–59; cultural difference and, 107; disregard of nation-state by, 167–69; education and, 48, 175–77; faces of, 166; hybridity and, 122–24, 126; identity and, 157, 160–63; interconnectedness of, 139; Native Americans and, 139; as neoliberalism, 136, 158; new struggles in, 169–72; personality of, 163–66; philosophy of praxis and, 130; possibilities of, 159–60; school reform and, 84, 94; solidarity and, 172–75; struggle and, 146; urban restructuring and, 152. *See also* neoliberalism
global North, 106
global South, 106, 164
Global Women's Strike, 172
Goals 2000: Educate America Act (1994), 85
Gramsci, Antonio, 61, 129, 147, 166
Guha, Ranajit, 40
Gutiérrez, Kris, 86, 113–14, 119

Habermas, J., 17
Hall, Stuart, 122
Harding, S., 53
Hardt, Michael, 132, 139–45, 146, 153–54, 169
Harris, Cheryl, 34
Hartsock, N., 52–53
Harvey, David, 81–82, 130–31, 137–38, 140, 153
Haymes, S. N., 91
Hegel, G. W. F., 138
hegemony, 130, 147, 167; of culture, 50, 55–59, 119
heterogeneity, 119–20
heteroglossia, 108, 113–15, 116, 119, 120, 126
historicism, 13–20
historicity, 9, 24, 27; liberation and, 10–13
history, 9–11, 13–14; as human activity, 11, 12; humanization and, 21–23
History and Class Consciousness (Lukács), 11
Holloway, J., 144
hope, 12, 13
Horkheimer, Max, 16
humanism, 132, 144; critical pedagogy and, 176; decentered, 30; Fanon on, 149, 150; materialism as, 46–48; rearticulation of, 21
humanity, 11
humanization, 9, 134, 173; ethical-political conception and, 146; history and, 21–23; oppressed, 11, 12–13; politics and, 16–18; redemption and, 13–14; resistance and, 25
human suffering, 144
Hurricane Katrina, 92

hybridity, 106, 107, 120–25, 126, 159
hyperauthoritarianism, 174
hypercommercialism, 19
hyper-disciplinarity, 87–89
hyperlogue, 20

identity/identities, 65, 176; border-crossing and, 121; collective, 124; critical postmodernist pedagogy, 112; culturally relevant pedagogy and, 110; democracy and, 115; diasporic, 120; global, 178; globalization and, 157, 160–63; heterogeneity and, 119–20; hybrid forms of, 106, 159; pedagogy of heteroglossia and, 114–15; race and, 72; social and political, 144; struggle and, 149
ideology, 116–117
immigrants, 73, 86, 135, 160; as service workers, 91
immigration, 133, 160
imperialism, 17, 32–33, 166, 180*n*5; cultural, 122; difference and, 106; movements against, 138; nationalism and, 117; neoimperialism and, 84, 98–101, 130, 164; neoliberalism and, 81; racist, 59; social justice and, 42; *vs.* terran class, 152
inclusion, 58
India, 40; seed control in, 164
Indian service workers, 162
indigenous communities, 58, 161
individuality, 170–71
industrialization, 167
information economy, 141–42
injustice, 130
Institute for Cultural Action, 136

interconnectedness, 139
international financial institutions, 81, 135, 136–37
Internet, 159
intersubjectivity, 113
invasions, 3, 17, 21, 32–33, 138

Jackson, George, 118
James, C. L. R., 25
Jameson, F., 20

Klein, Naomi, 176
knowledge/power relationship, 117–18

labor, 31, 63, 94, 97, 99; army of, 170; capital and, 138, 145; cheap, 133, 167; class and, 95; creativity and, 140; international division of, 164; labor movements and, 131; labor process and, 163–64; of women, 131, 160–61, 172
Ladson-Billings, Gloria, 109, 110
language, 56, 106, 148–49; authoritarian, 120; of class struggle, 133; cultural difference and, 107; educational politics and, 126; English, 73, 85, 108; hybrid, 121; knowledge/power relationship and, 117–18; pedagogy of heteroglossia and, 113, 114–15; as political, 120, 126; for voices of students, 112
Latin America, 133
Latino/a youth, 87
law enforcement, 87, 99
leadership, 39–42. *See also* authority
learning, 12, 117; contemporary schooling and, 79; gender and, 64–66; multitude concept of, 143; pedagogy of heteroglossia

and, 113; progressive, 105; separation from, 98; standardized testing and, 19, 29, 60, 79, 85, 164; through struggle, 46. *See also* students; teaching

"Letter to a Frenchman" (Fanon), 38

liberation, 24, 44; Freire on, 61, 134, 154; historicity and, 10–13; imagination and, 175–77; movements for, 39–42, 43, 105–6

liberatory praxis, 10–11

Lipman, Pauline, 90, 152

liquid modernity, 169–70

literacy, 113; cyberliteracy and, 175; Freire on, 134; as political, 120

local struggles, 152, 154

Lorde, Audre, 24–25

low-income people, 136–37

low-income students, 88, 100, 102

Lukács, Georg, 11, 52, 73

Luxemburg, Rosa, 46, 143

Lyons, S. R., 88

marginalization, 99, 101, 102; exploitation and, 138; globalization and, 160; oppressor/oppressed paradigm and, 136

markets, 122, 168

Marx, Karl, 36, 80–81, 94–98, 102–3, 138; on capitalism as progress, 163; on personality of capitalism, 165, 182*n*2; on primitive accumulation, 81–84; on species-being, 150; on struggle against capitalism, 159; on surplus population, 89

Marxism, 25, 31, 52; capitalist accumulation and, 59; Fanon on, 30, 72; Freire and, 12; globalization and, 130; materialism and, 35, 47–48

masculinist worldview, 52

materialism, 33–39, 59, 71, 73, 150, 177; dialogue and, 120; historical, 61–62; humanism as, 46–48; oppression and, 74; standpoint theory and, 51–54

mathematics, 64

McChesney, R. W., 19

McClintock, Anne, 72

McLaren, Peter, 50, 59–63, 68, 70, 73, 74

McNeil, L. M., 19

media, 45

men, 172; boys, 64, 66. *See also* gender; women

Mexico, 161

middle class, 162

Mies, M., 161

migrations, 133

militarization, 101; of schooling, 32, 46, 88, 100

military aggression, 130

modernist paradigm, 111

modernization *vs.* development, 14–15

Mohanty, Chandra, 72–73, 131, 144, 147, 153

movements of resistance, 66, 154; antiwar, 140, 142, 166; global, 130–31, 152

multiculturalism, 34, 56, 58, 76, 105, 120–21; capitalist culture and, 125; critical, 126; global South and, 106

multitude, 132, 153–54, 169; multitude/empire paradigm and, 139–45. *See also* empire

Multitude (Hardt and Negri), 139

national boundaries, 167
nationalism, 117, 161; ethnic movements for, 120–25
nationalist movement (India), 40
Nation at Risk, A (National Commission on Excellence in Education), 85
nation-state, 167–69
Native Americans, 58, 102, 139
"natural" learning, 64–65, 74
negation, 37; of oppression, 144, 149
Negri, Antonio, 132, 139–45, 146, 153–54, 169
neocolonialism, 15, 35, 75; education and, 72; Freire on, 135; oppression and, 76–77
neoimperialism, 84, 98–101, 130, 164
neoliberalism, 41–42, 59, 81–82, 170; deterritorializations of, 146; dispossession and, 138; double-sided movement of, 174; education and, 48; globalization and, 136, 158; global movements against, 123, 130; Native American movements and, 139; overcoming, 43; overproduction and, 167; racism of, 91; school reform and, 94; social justice movements and, 171; state power and, 101; surplus population and, 89; *vs.* terran class, 152
New Orleans, Louisiana, 92
Nieto, Sonia, 50, 55–59, 70, 73, 74
No Child Left Behind Act (2001), 32, 69, 85, 88, 92
non-governmental organizations (NGOs), 151

North American Free Trade Agreement (NAFTA), 161

occupation, 17, 18–20, 118
opposition, 123, 147, 159; class politics and, 137–38; collective subject of, 171; democratic, 140; educational, 175; fabric of, 176–77; global, 126, 132, 153; hybridity and, 122; praxis for, 140, 145; subjects of, 142; transnational, 125, 126
oppressed, the, 165; global class of, 150–51; humanization of, 11, 12–13; political agency of, 24; social psychology and, 14–15; subjectivity of, 40–41. *See also* Freire, Paulo
oppression, 27, 49–77, 147, 173; as capitalist accumulation, 50, 59–63; compound standpoint of, 70–73, 73–76; contemporary, 21; convergence of resistance to, 67–68; as cultural hegemony, 50, 55–59; dialogue and, 22–23; as discursive effect, 50, 63–67; economy of, 43; heterogeneity and, 119; hope and, 12; intersecting experiences of, 69–70, 71; language of class and, 149; left analysis of, 33; negation of, 144, 149; politics and, 17–18; postmodernism and, 111; redemption and, 13–14; reinvention of Freire and, 10; standpoint theory and, 51–54. *See also* Freire, Paulo
oppressor/oppressed paradigm, 131, 132–39, 154; multitude/empire paradigm and, 140

organization, 142–43; cellular form of, 151; solidity of, 177; of violence, 164
otherness, 111
outrage, 145–46
overproduction, 166–67

patents, 137
patriarchy, 62, 64, 74–75, 76; critical postmodernism and, 111; teaching against, 121
patriotism, 32
pedagogies of difference, 108–25; critical postmodernist pedagogy, 108, 110–13, 114, 116; culturally relevant pedagogy, 108–10; discourse and power and, 115–20; heteroglossia and, 113–15; hybridity and, 120–25
pedagogy: culturally relevant, 107, 108–10, 114, 116; democratic, 44; dialogue and, 26; of heteroglossia, 108, 113–15, 116; liberatory imagination and, 175–77; solidarity and, 27, 72, 172–75; standardization of, 19–20. *See also* education; schools/schooling; teaching
Pedagogy of the Oppressed (Freire), 10, 15, 18, 138
personality of globalization, 163–66
philosophy of praxis, 129
plunder, 82, 83, 102; overproduction and, 167; primitive accumulation and, 80; spiritual, 138, 168
political organization, 151
politics, 27, 106, 116, 130, 144; colonialism and, 72; dialogue and, 22–23, 42–46; economy of, 62–63, 73, 123; educational, 126; Fanon and, 29; Freire and, 16–18; global, 142; hybrid forms of, 159; internal/external contradictions of, 39–42; language and, 120, 126; of location, 131; of transcendence, 147; pedagogy of heteroglossia and, 113; radical, 107
poor people, 136–37
poors, the, 162
poor students, 88, 100, 102
Popular Assembly of the Peoples of Oaxaca, 152
postmodernism, 20, 30, 111
poststructuralism, 63, 64–67, 106; feminist, 76
power, 114, 145; as antagonistic, 151; biopower and, 141; children of elites and, 119; critical reading of, 173; discourse and, 115–20; flexibility of, 153; Freire's analysis of, 133; global, 166–69; as indifferent, 164; internal crises of, 130; multiculturalism's challenge to, 125; new dimensions of, 166; oppressor/oppressed paradigm and, 134, 154; resistance and, 148; students' relationships to, 172–73; violence and, 17
praxis, 129, 177; globalization and, 130; multitude/empire paradigm and, 139–45; open-ended, 43; terran class and, 150–53
prejudice, 75
primitive accumulation, 80, 81–84, 137, 180*n*3, 182*n*3; contemporary schooling and, 94, 95; empire and, 140; in U.S.,

102–3; marginalization and, 101; methods of, 96; separation of producers from production means by, 97; technology of solidarity and, 163
principle of difference, 105
prison system, 4, 99, 101; detention as, 88
privatization, 79, 96, 167; of basic services, 81; of communal land, 82–83; of education, 60, 91–94, 139; global movements against, 123
privilege, 74, 75; global consciousness and, 125; preservation of, 100, 113; in schools, 57, 69, 86, 113
problem-posing education, 10, 16, 22, 25–27, 135
progressive movements, 133
proletarianization, 167
proletarians, 97, 98; violence toward, 100
protests, 55, 135, 143, 144, 160, 162, 164; antiwar, 166; global, 140, 142, 151; youth in, 175
public life, 174
public resources, 81, 82–83, 84, 137, 167
public schools. *See* schools/schooling
public sphere, 167

queer identities, 57

race, 67, 75, 79; accountability initiatives and, 118; class and, 31, 39, 73; critical postmodernist pedagogy and, 111; hyper-disciplinarity and, 87–89; identity and, 72; pedagogy of heteroglossia, 114; students oppressed by, 68; territory and, 58; whiteness and, 31, 34, 71. *See also* Black and brown bodies; class; culture; gender; students of color
racism, 29, 39, 69; Americanness and, 161; assimilationist, 107; colonialism and, 30–33, 71; critical postmodernism and, 111; cultural hegemony and, 55; culturally relevant pedagogy and, 110; Fanon on, 30–33; neoimperialism and, 100; neoliberalism and, 91; oppressor/oppressed paradigm and, 134; in schools, 29, 32, 34–35, 38, 48, 55; segregation and, 75, 90; social justice and, 42; success and failure and, 38; systematic, 37, 55–59, 74, 102, 116
radio, 45
reality, 11–12, 116, 176
relationships, 154; across boundaries, 120; democracy and, 153; impact of globalizing processes on, 157; students' to power and exploitation, 172–73
religion, 103
reorganization of social life, 2, 123–24, 143, 159–60, 164–65, 169; primitive accumulation and, 81
resegregation, 89–91
resistance, 24–25, 148, 154, 159, 178; by indigenous Mexicans, 161; convergence of modes of, 67–68; economy of, 43; to globalization, 146; global movement of, 130–31, 152; identity and, 162; multitude's

distance from, 143; participation in, 177; in U.S., 61–62
resources, 35, 72; communal, 81, 82–83, 84, 137
retention policies, 88
retrograde motion, 23–25
revolution, 13, 16, 22, 52–53, 142, 144–45
Robinson, William I., 59

Said, Edward, 15, 145, 180n5
Saltman, K. J., 81–82, 92
Sandoval, Chela, 69
Sassen, Saskia, 26, 91, 160
school reform, 84–94; hyper-disciplinarity of, 87–89; privatization of, 91–94; racial inequity of, 89–91; scripted curricula and, 85–87
schools/schooling, 52–53, 75, 79; charter schools and, 86, 92, 93–94, 97; choice of, 91–94; clearings in, 94, 98–101; corporatization in, 94–98; enclosures in, 91, 94–98; militarization of, 32, 46, 88, 100; political reconfiguration of, 63; privatization of, 60, 91–94, 139; as punishment, 174; racism in, 29, 32, 34–35, 38, 48, 55; reorganization of, 82; separation of producers from means of production in, 97
scripted curricula, 19, 20, 55, 60, 75, 85–87
seed, 164
segregation, 29, 75–76; desegregation, 62; resegregation, 89–91. *See also* racism
separation of producers from means of production, 97–98, 102

service economy, 91, 141–42, 160–61
sexism, 57, 62, 69, 134
sexuality, 131
shared histories, 120
situatedness, 132
Sklair, Leslie, 150
slavery, 102
social crises, 124
socialism, 168–69
social justice, 26, 42, 61, 85, 159–60, 171; teaching and, 160
social life, 93, 106, 129, 153, 178; as alien to individuals, 164–65; capital penetration into, 60, 96, 123, 167, 168, 170; communication and, 169; contradictions in, 148; fluid, 171; identity and, 144; inequalities in, 55, 65; materiality of, 35, 48, 71; multitude and, 140, 141; multitude/empire paradigm and, 141–42; politics of discourse and, 116; prison system and, 4, 101; psychology of, 14–16; regulation of, 99; reorganization of, 2, 81, 123–124, 143, 157, 159–60, 164–65, 169; reproduction of, 64–65, 148, 149, 164, 170; standpoint theory and, 53; success/failure and, 37; textuality of, 120, 126; violence of, 125
social movements, 172, 175
social services, 60, 101, 168, 170
social transformation, 172
solidarity, 47, 48, 112, 153, 177; against elites, 162; feminist, 131; hybridity and, 107, 121, 123, 124; multitude and, 144;

nationalism and, 117; oppositional, 147; pedagogy and, 27, 72, 172–75; problem of, 125–27; technology of, 163; terran class and, 154; working class and, 71
South Africa, 133, 137, 162
sovereignty, 139–41; global, 132
Spivak, G. C., 16
standardized testing, 19, 29, 60, 85, 164; gentrification and, 79
standpoint theory, 51–54; compound, 70–73, 73–76; limitations of, 68–70
state power, 101, 140, 142, 145
structural adjustment, 139; of education, 60, 97–98, 152; of social services, 168; of welfare, 101
struggle(s), 46, 61, 146, 176, 178, 180*n*2; anticolonial, 15, 72; class, 131, 133, 138, 141; against capitalism, 159; diversity of, 153–54; identity and, 149; individual, 153; new, 169–72; objective and subjective dialectics of, 145; social contradictions and, 148; solidarity and, 162, 173; terrains of, 145; of terran class, 151; total experience of, 143
students, 68; agency of, 119, 176; as consumers, 97; culture of, 109; relationship to power and exploitation of, 172–73; as teachers, 163, 176; transnational oppositional movements and, 126; voices of, 111–12, 113, 117. *See also* teachers
students of color, 32, 75, 76, 87, 90; achievement of, 107; culturally relevant pedagogy and, 109–110; discipline and, 99; expectations for, 37; marginalization of, 102; military recruitment of, 100; needs of, 41; poststructural analysis and, 67; power relations between white teachers and, 114; resources for, 35
subalterns, 39–42
subjectivity, 40–41, 63, 170
success, 37–39, 49; discouragement of, 89
suffering, 144
surplus capacity, 167
surplus population, 89, 98
surplus production, 148
surplus value, 141
surveillance, 88–89
systematic racism, 37, 55–59, 74, 102, 116

teachers, 19, 88, 114, 116, 173; dialogue and, 44, 45–46; education for, 176; humanism and, 48; as learners, 163, 176; radical, 66; voices of, 115; as wage-laborers, 97. *See also* students
teaching, 26, 44, 79, 117; as collaborative, 22, 173–74; culturally relevant pedagogy and, 110; against patriarchy, 121; progressive, 105; for radical change, 121; for social justice, 160; solidarity and, 27, 72, 172–75; test preparation and, 85. *See also* learning
technology of solidarity, 163
telecommunications, 159
terran class, 145, 149–54, 182*n*5; identification of, 154; praxis and, 150–53; struggles of, 151

territorialization, 132, 135, 154
territory, 58; global, 146
testing, high-stakes, 98
test preparation, 85. *See also* standardized testing
test production and scoring industry, 86, 87
totality, 161
tracking, 79; retracking and, 89–91
trade unions, 138
transformative pedagogy, 107, 176
transnationalism, 60
transnational protest movements, 126, 160
trends in contemporary education, 79, 81–84, 102–3
truth, 47

uncertainties, 169
United States: apartheid system in, 90; government accountability in, 164; neoimperialism of, 100; primitive accumulation in, 102–3; resistance in, 61–62
urban restructuring, 152
utilities, 137

vagabonds, 98–99
violence, 17, 21, 76, 118, 125; of capital, 162; capitalist exploitation and, 33, 103; colonialism and, 31, 71; cultural and economic, 136; educational, 32; Fanon on, 29; history of, 150; humanism and, 47; against indigenous peoples, 58; language and, 73; materiality of historical, 74; of neoliberal deterritorializations, 146; organizers of, 164; primitive accumulation and, 80, 84; toward proletarians, 100; school reform and, 84; in seizure of pubic resources, 81; to social difference, 142; of state and capital, 145
voices, 20; of students, 111–12, 113, 117; of teachers, 115
voucher schemes, 86, 92, 97
Vygotsky, L. S., 113

wage-laborers, 82, 83, 97, 99
wages, 89
Walkerdine, Valerie, 50, 63–67, 68, 70, 74
war: antiwar protests and, 140, 142, 166; biopower and, 140–41; global movements against, 123; on terror, 18
wealth, 30, 31, 49; children of elites and, 119; in colonialism, 71; primitive accumulation and, 80, 82, 83, 84; of U.S. capitalism, 102–103
welfare, 60, 101, 170
white population, 59, 90, 108. *See also* race
white students, 32, 90. *See also* students of color
white supremacy, 43, 55, 67; Americanness and, 161
white women, 69
women, 64, 66, 69–70, 76; Black, 53; feminism and, 52–53, 70; labor of, 131, 160–61, 172; service work and, 160–61; white, 69; women's movements and, 131. *See also* gender; girls
working class, 68; learning and, 46; revolution and, 52; solidarity and, 71; struggles of, 131, 138, 141
World Social Forum, 140, 171
World Trade Organization, 164

Wretched of the Earth, The (Fanon), 40, 45, 149

youth, 159–60, 172, 174, 175; criminalization of, 44; of color, 87–88; teachers' collaboration with, 173. *See also* students; students of color

Žižek, Slavoj, 124